TOM PAINE'S IRON BRIDGE

TOM PAINE'S IRON BRIDGE

A slight Sketch of Tho.ˢ Paine's patent cast Iron Bridge proposed to be

A. The front of the House at the North end of Green's
D. The top of the great South Rock 35 feet above the low
G.G Two Cylinders. H. The North Key. I. The great
the high-water level 70 feet.

BUILDING A UNITED STATES

over the River Wear near Sunderland. *Scale 20 feet to one Inch.*

The road along the Banks. *C. The level of the Valley.*
evel. *E. The high water level.* *F. The low water level.*
ch, Cord 200 feet; versed sine 25 feet; height above

Edward G. Gray

W. W. NORTON & COMPANY
Independent Publishers Since 1923
New York   London

For information about permission to
reproduce selections from this book, write to
Permissions, W. W. Norton & Company, Inc.,
500 Fifth Avenue, New York, NY 10110

For information about special discounts for bulk
purchases, please contact W. W. Norton Special Sales at
specialsales@wwnorton.com or 800-233-4830

Manufacturing by Quad Graphics, Fairfield
Book design by Helene Berinsky
Production manager: Louise Mattarelliano

ISBN 978-0-393-24178-5

W. W. Norton & Company, Inc.
500 Fifth Avenue, New York, N.Y. 10110
www.wwnorton.com

W. W. Norton & Company Ltd.
Castle House, 75/76 Wells Street, London W1T 3QT

1 2 3 4 5 6 7 8 9 0

For Sophie Elena Gray and Tobias Rutledge Gray

Contents

The mechanic should sit down among levers, screws, wedges, wheels, & c. like a poet among the letters of the alphabet, considering them as the exhibition of his thoughts; in which a new arrangement transmits a new idea to the world.

ROBERT FULTON,
A Treatise on the Improvement
of Canal Navigation (1796)

Author's Note:
On Architects and Engineers

In keeping with the age in which Paine lived, I have elected to refer to him and his fellow bridge builders as "architects" and to refer to their bridge-making activities as "architectural" activities.

The terms "architect" and "engineer" now denote two distinct, though related, provinces of knowledge. The architect tends to be responsible for form, for aesthetic choices, for the parts of buildings we take note of. The engineer tends to be associated with the parts we do not see—the underlying skeletons of skyscrapers or the trusses that hold up a stadium roof. It is the architect who designs the structure; the engineer who makes it stand.

In the eighteenth century, this distinction did not exist. Those who imagined buildings also accounted for their structural worthiness. Indeed, in many cases they also built the buildings. Much that we would now leave to architects and engineers was done by carpenters, bricklayers, and stonemasons. A combination of experience, intuition, and widely available builders' handbooks guided their aesthetic and structural choices. Insofar as the term "architect" appeared in English, then, it as often referred to builders as it did designers of buildings and their inner workings.

What was true of buildings was also true of bridges. For the most part, these structures, whether stone or timber, were designed and built by the same people, stonemasons or carpenters. Insofar as there was a single English-language term designating the designers and builders of structures, it was architect.

Outside of France, the term "engineer," derived in its modern usage from the French *ingenieur*, generally had military associations. Engineers were those who built and operated "engines" or devices—catapults, assault towers, bridges, and later artillery—used to breach castle walls. In the eighteenth century, the meaning had expanded to include those who built roads, fortresses, and other military installations. But the term would begin to be widely applied to the builders of civil, nonmilitary structures only in the early nineteenth century.

TOM PAINE'S IRON BRIDGE

Introduction

ON A MILD SEPTEMBER DAY in 1790, Thomas Paine, the American revolutionary, stood before a 110-foot iron-arch bridge on Lisson Green, at the corner of Edgware Road and the new Marylebone Road on the outskirts of London. Over the course of the previous few months, Paine had overseen a crew of five workmen as they erected the structure out of cast-iron parts shipped from Yorkshire. It was a strange sight, a bridge spanning no waterway or gorge. But it represented a compelling vision, Paine's plan for the enduring success of the new American republic.

Nothing remains of Paine's bridge, and the Green has long since disappeared beneath council houses and the Marylebone Flyover, an elevated extension of the A40 roadway. You can hardly imagine that the place was once a way station for weary travelers or that it served as a transit point for herdsman moving sheep, pigs, and cattle east, along Marylebone Road (at the time known as New Road) to the Smithfield Market. But in 1790, the Green was one of the few places near central London suited to the display of an architectural marvel. Much of the city remained a warren of medieval roads, and although developers had begun incorporating open squares in Mayfair, Bloomsbury, and other Georgian neighborhoods, there was little chance the landowners and their genteel tenants would tolerate public spectacles undertaken by foreign visitors.

Lisson Green was a different kind of space. Some years before Paine began building his bridge, it had been the location of a white-paint manufactory, an enterprise whose emissions were rivaled only by the leather tanners' for foulness. Now, with the ebb and flow of travelers, and its nearby entertainments, including the Yorkshire Stingo Pub and Lord's Cricket Grounds, Lisson Green was ideally situated to display a new industrial-age marvel.[1]

Paine's bridge was composed of hundreds of cast-iron bars fastened together to form a series of five parallel arches, the tops of which were linked by heavy wooden planks. The structure loomed above the ground like a whale's arched back situated between two wooden platforms, which countered the downward force of the arch. At a distance, the low-slung structure—rising to a height of five feet at its peak—might have seemed part of the landscape. It emerged from the ground, as if a small segment of a full circle. The visible lattice of iron bars, joined together to form the structural arches, and the pedestrian railings were all that interrupted the pleasing geometry of the bridge.

For strollers on the Green, the purpose of the bridge was entirely ornamental. It crossed no river or stream. Nonetheless, through September of 1790, a steady stream of Londoners paid one shilling to traverse the bridge, walking its gentle incline to a rounded peak. From here, they gazed upon the Green and the tavern across the way. The easy descent of the bridge deposited visitors on the Green with the feeling of having barely left the ground.

Such was, in fact, the intent of its architect. In the eighteenth century, movement could rarely be described with adjectives like "gentle" and "easy." Whether traveling by sea, inland waterway, or land, the risks to person and property were countless. This was especially true for travelers in Thomas Paine's adopted home of Pennsylvania. The state's hills, from the Alleghenies east, fed a river system that ultimately emptied into the Chesapeake and Delaware Bays. In summer and fall, those rivers were vital conduits for goods being shipped to Baltimore and Philadelphia. But in winter and spring, they became impassible torrents of water and ice.

Having established his reputation as the American Revolution's most celebrated propagandist, Paine now hoped to free his fellow Pennsylvanians from these riverine hazards. Once his bridge was endorsed and adopted in the country of his birth, he would return to America and throw his span across the Schuylkill River at Philadelphia. That bridge would give rise to imitators, which would transform the Pennsylvania countryside and, ultimately, the whole of the United States from a welter of natural obstacles and commercial interruptions into a unified empire of liberty.

OF THE MANY ESSAYS Thomas Paine wrote, among the least known is "The Construction of Iron Bridges." This brief history of Paine's architectural career, written in 1803, was of no particular interest to his political followers, nor has it been to his many subsequent biographers. The essay after all has little to do with the radical critique of hereditary monarchy or the cult of natural rights for which Paine has been so justly celebrated. But it is a window into his world. Many of the luminaries in Paine's circle were inventors. Paine's friend Benjamin Franklin devised bifocals, the lightning rod, the glass armonica, and countless other devices. Another friend, Thomas Jefferson, invented an improved plow and a mechanism for copying letters. Some revolutionary leaders not known for their inventions devoted time to building things. George Washington often seems to have lavished as much attention on his house at Mount Vernon as on matters of state. From this vantage, Paine seems no different.

But Paine was different. Unlike so many of his American contemporaries, Paine had a narrow field of interests. He never showed any passion for art or philosophy. He claimed repeatedly to have learned little from books. He did have other mechanical interests. He attempted to invent a smokeless candle and later in life he contemplated a perpetual-motion machine driven by gunpowder. But neither of these consumed Paine in the way his bridge did. Indeed, far from a gentlemanly hobby, bridge architecture became a career for Paine. In his essay on iron bridges, he wrote that he had had every intention of devoting himself fully to architecture but was drawn away by events beyond his control.

The most disruptive of these was the 1790 publication by the British politician, and former friend of Paine, Edmund Burke, of *Reflections on the Revolution in France*. For Paine, Burke's fierce denunciation of the course of events across the English Channel was about much more than France and its revolution; it was an attack on the political ideals on which his adopted country had been founded and on which a just future would depend. "The publication of this work of Mr. Burke," Paine explained, "absurd in its principles and outrageous in its manner, drew me . . . from my bridge operations, and my time became employed in defending a system then established and operating in America and which I wished to see peaceably adopted in Europe." The refutation of Burke became "more necessary," for the moment, than the construction of the bridge.

Paine's response to Burke, *Rights of Man*, the first part of which appeared in early 1791, earned him the admiration of a broad coalition of Atlantic radicals, from the United States to Ireland, England, France, and across much of the rest of Europe. Paine's rejection of Burke's antirevolutionary doctrine was so popular in France that in the early fall of 1792, after the appearance of the second part of *Rights of Man*, Paine was elected to serve in a new French constitutional convention.[2]

Paine made it clear to his correspondents that, even after fleeing England and enduring much revolutionary chaos in France, he intended to return to the United States to bridge the Schuylkill with an arch of iron. But his plans were derailed. The Wars of the French Revolution and America's own reluctance to receive Paine made safe passage impossible. When Paine finally did return to the United States, in 1802, he found no supporters for his bridge and was forced to abandon architecture once and for all.[3]

IN SOME WAYS, PAINE'S architectural career is indicative of so much that has come to be known about him. Although he was among the American Revolution's most vocal and radical proponents, and author of some of the most influential political pamphlets ever written, so much of what Paine stood for came to naught in his lifetime. His dream of an irresistible wave of democracy, spreading from America to Europe, and ultimately enveloping the world, came crashing down amid the wreckage of the French Revolution. His expectation that government by the people would bring an end to war, monarchy's chosen tool of statecraft, was proven hopelessly naïve. His hatred of the most brutal and controversial labor system of his age, American chattel slavery, was met with only the most incremental adjustment of attitudes and law. His ambitious programs to alleviate human misery through more equitable systems of taxation would come nowhere near realization for decades, if not centuries. Like so many radical visionaries, Paine often seems more the man of ideas than of action, more the dreamer than the doer.

The perception is not just the work of hindsight. Many of Paine's contemporaries regarded him as a hyperventilating crank. Gouverneur Morris, the New York attorney and statesman, dismissed Paine as "a mere adventurer *from England*, without fortune, without family or connections, ignorant even of grammar." The idea that such a person could be responsible for some of the age's most eloquent political statements was too much for some of Paine's contemporaries to accept. John Adams, who came to despise Paine in the years after the French Revolution, wrote of him that "there can be no severer a satyr on the age. For such a mongrel between pig and puppy, begotten by a wild boar on a bitch wolf, never before in any age of the world was suffered by the poltroonery of mankind, to run through such a career of mischief." Paine's enemies even attacked his architectural ideas. During a 1788 trip to Philadelphia, the liberal French politician Jacques-Pierre Brissot de Warville was told that not only was it "generally agreed" that Paine had plagiarized portions of his war-time cri de coeur, a series of essays entitled *The American Crisis*, but "he is also accused of having copied the plan of his iron bridge."[4]

Paine himself had much to do with this hostility. He was flighty and impolitic, prone to drunken fits and debilitating personal grudges. Unlike so many of his equally well-known contemporaries, he never seemed to grasp the fundamental social truth of his age, namely that to gain the good graces of the powerful, it was necessary to flatter their sense of propriety and social superiority. At times, the democratic Paine was flagrantly oblivious to this mandate, all too prepared to arouse the fury of those very elite gentlemen who could do much to advance his causes.

In late 1778, during one of the most precarious moments of the American Revolutionary War, Paine publicly accused the American agent and merchant Silas Deane of improperly profiting from the sale of French arms and munitions. This kind of principled attack, so typical of Paine, would likely be little remembered but for its dangerous repercussions. In exposing Deane's profiteering, Paine revealed a covert French alliance begun well before the formal Franco-American alliance of February 1778. The revelation compromised earlier diplomacy and became a terrible embarrassment for the French government. To Paine's American detractors, the slip seemed to confirm his duplicity. Surely a sincere champion of American independence would never so clumsily imperil relations with America's most important ally. Even to Paine's friends, this kind of political ineptitude meant that when it came to delicate matters of state, he simply could not be trusted.

But he was to be reckoned with. John Adams confessed that his own age might just as well have been called "the age of Paine." For better or worse, the immigrant Englishman, the son of a humble stay-maker, and the notorious pamphleteer would leave his mark. Paine's begrudging contemporaries had to acknowledge that his writings struck a chord. *Common Sense*, Paine's call for American independence from Britain, which appeared in 1776, sold tens of thousands of copies. *Rights of Man*, a dense work of politics and economics, sold thousands of copies, far outselling Burke's *Reflections*. Had it not been for a deliberate British campaign to suppress the pamphlet, many tens of thousands more would surely have been sold.

Paine's language resonated in his own age as it does in ours. From *Common Sense*: "We have it in our power to begin the world over again." From *The American Crisis* of December 23, 1776: "These are the times that try men's souls. The summer soldier and the sunshine patriot will, in this crisis, shrink from the service of their country; but he that stands it *now*, deserves the love and thanks of man and woman." Thomas Jefferson described Paine's gifts best when he wrote that "no writer has exceeded Paine in ease and familiarity of style, in perspicuity of expression, happiness of elucidation, and in simple and unassuming language."[5] Paine wrote for the masses and he did so with success known by no other political writer until, perhaps, Karl Marx.

To his many critics, Paine's popularity merely confirmed his reputation as a revolutionary gadfly. It is all good and well to assail an old world of kings and tyrants, to lament the failings of the British government, and to deride the arrogance and superficiality of the rich and powerful. But where was Paine, his detractors always asked, when the time came to build the new world so vividly imagined in his writings? As John Adams said of Paine, and as Paine's career as pamphleteer might suggest, he had "a better hand at pulling down than building."[6]

I HAVE COME to know a different Thomas Paine. The Paine I know was as committed to building a new world as to tearing down an old one. This Thomas Paine emerged only after I came to see that his political thought and his architecture were of a piece. They reflected the same capacious revolutionary ideology. It took me some time to find this Paine. After years of reading his writings and following his life story, I began to wonder why, at the height of his literary powers, Paine turned to architecture. What I came to see was that this turn was not an abrupt life change for Paine. It was a logical step forward for a man committed to the causes of democracy and liberty.

Free societies, Paine believed, would work only insofar as their citizens could communicate with one another. In the United States, this assumption became a source of intense political disagreement during the debates over a new federal constitution in 1787 and 1788. Opponents of the Constitution doubted that so vast a country could bind together its disparate parts. Free inhabitants of a large territory would inevitably seek to govern themselves without the intrusions of distant authorities. As one anonymous opponent of the Constitution wrote, "History furnishes no example of a free republic, anything like the extent of the United States."[7] The only way America's far-flung parts could be bound together into a single political community was with the instruments of tyranny. But even these would ultimately fail, as the examples of the Roman Empire and Britain's own empire in America made all too plain.

For James Madison, the political visionary behind the new constitution, an American empire would be different. Infrastructure and communications would make size irrelevant. "The intercourse throughout the union," he assured skeptics, "will be daily facilitated by new improvements," by which Madison meant bridges, roads, canals, dredged rivers, and other arteries of connection and commerce. As long as Americans could overcome barriers to communications, whether the natural ones of threatening weather and rushing rivers or the human-made ones of faction and party, there was every reason to be optimistic about the prospects of even so sprawling a republic as the United States. These are precisely the ideals that compelled Paine to devote himself to something as prosaic as an iron bridge.[8]

In Paine's lifetime, these ideals went largely unrealized. The first permanent bridge would be built across the Schuylkill before he died, but it was a bridge that entirely ignored the architectural principles Paine advocated. The National Road and the Erie Canal, the early American republic's two greatest internal improvements, were begun in 1811 and 1817, respectively, well after Paine's death. These extraordinary public works, along with hundreds of others less well-known, drew the country together in a way that even Paine could never have imagined. Time and again, Europeans traveling in the United States marveled at the nation's capacity to create new arteries of commerce and communications. Alexis de Tocqueville, the celebrated French student of American democracy, was stunned by all the roads, bridges, and canals he saw while traveling through the country in 1831. Regarding these "means of carrying rapidly from place to place the produce of industry and of thought," he remarked,

> I do not pretend to have made the discovery that they served the prosperity of a people. That's a truth universally felt and recognized. I say only that America makes you put your finger on this truth, that it throws the fact more in relief than any other country in the world, and that it is impossible to travel through the union without becoming convinced, not through argument but by the witness of all the senses, that the most powerful, infallible way of increasing the prosperity of a country is to favor by all possible means a free intercourse among its inhabitants.[9]

Tocqueville's sense of American prosperity may have been exaggerated, but his sense of America's capacity to draw together its widely scattered continental populace was not. America's internal improvements were astonishing even to this Frenchman, whose country was celebrated the world over for the quality and extent of its roads and bridges. And his conclusion that that capacity was a foundation of American democracy would have entirely confirmed an ideal Paine championed decades earlier.

Paine's faith in the unifying powers of internal improvements reflected the optimism of his revolutionary age. He never grasped, as Tocqueville did decades later, that even with these advances the United States could become deeply and catastrophically divided. Paine understood sectional division in a more local sense. It was an issue for the state of Pennsylvania, which had been divided since the colonial era between an alienated and underrepresented western population and an older, more prominent eastern one. In drawing together the state's sections, Paine believed, iron bridges would do much to address these divisions. A similar process, he assumed, would follow in other states. Here Paine's thinking bears a utopian cast. But, as I hope the following pages make clear, that utopianism is really only evident with hindsight. Paine's own perception was that he was offering a concrete, practical solution to a serious political threat.

As a product of the European enlightenment and its neoclassical habits of mind, Thomas Paine admired symmetry. Whether in governments or bridges, balance and rational order were the ultimate tests of worth. Paine's bridge would have all the grace and balance of a small segment of a large circle. Government at its best would express a similar aesthetic balance. Its size and complexity would be scaled to the needs of the governed, not the interests of officeholders. To Paine, the ultimate insult to this natural law of politics was the British empire, a governing instrument whose form defied all reason. As Paine remarked in *Common Sense*, "there is something absurd, in supposing a Continent to be perpetually governed by an island. In no instance hath nature made the satellite larger than its primary planet."[10]

An alternative political order was being devised in Pennsylvania and its capital city, Philadelphia. In this new American republic, empire would take a felicitous turn toward symmetry and balance. And it would do so, Thomas Paine came to believe, with the help of his remarkable new invention.

1

River City

In the early fall of 1774, as the *London Packet* sailed through the mouth of Delaware Bay and up the river toward Philadelphia, its 120 passengers lay about the vessel, weak and thin. Their passage to America had been brutal. Shortly after leaving England, a "putrid fever" swept through the ship. The illness ravaged passengers and crew alike, killing five and leaving the ship's English captain, John Cooke, fighting for his life. Fortunately a doctor was aboard; he was able to administer the array of unctions, reassurances, and spirits that were the staples of eighteenth-century medicine.

Among the physician's patients was the thirty-seven-year-old Thomas Paine. Paine's case was particularly severe. After the *London Packet* docked at the Philadelphia wharf, he continued to waft in and out of a feverish delirium, unable even to raise himself long enough to traverse the gangway from ship to shore. A Philadelphia doctor took charge of Paine's care and made arrangements for him to be taken to the home of relatives of Captain Cooke. There, Paine would spend the next six weeks convalescing, slowly acquiring a sense that he had arrived in America.[1]

Paine was not typical of the passengers aboard the *London Packet*. Most were indentured servants, many traveling with families, bound by legal agreements to repay the cost of transit with as many as six or seven years labor. The vast bulk of that labor would take place on the small farms that dotted the Delaware River valley, or the rapidly expanding counties to the south and west, across Philadelphia's other river, the Schuylkill.

Paine came as a free man. He was beholden to no one, with the possible exception of his physicians and the great Benjamin Franklin, whom he had met in London and whose letters of introduction would ease Paine's arrival in America. He was also a man of means, or at least of sufficient means to avoid the kind of indebtedness that most likely landed many of his fellow passengers in servitude.

PAINE WAS BORN in 1737 into England's growing middling population. His father, a Quaker stay-maker, and his mother, the Anglican daughter of an attorney and town clerk, were able to provide a basic education for their only child (Paine's younger sister, Elizabeth, died as an infant). But by age twelve, Paine was forced to start working, first as an apprentice to his father and subsequently as a common seaman aboard the British privateer *King of Prussia*. Life at sea did not agree with the young Paine and he returned to stay making, establishing a shop in Sandwich, Kent. The twenty-two-year-old Paine also married Mary Lambert, but a year after their marriage, Mary died in childbirth and Paine's business began to fail.

Looking beyond the family trade, Paine seized on a new career path. After Mary's death, he returned to his parents' home in Thetford, Norfolk, to prepare for entry into the Excise Service, the government agency charged with collecting taxes on beer, spirits, wine, leather, paper, soap, candles, and other taxable goods. Following the path of his father-in-law, James Lambert, Paine would spend over a year seeking to become one of Britain's more than 2700 excise officers.

Preparations for the Excise was, in some ways, not unlike those for other eighteenth-century professions, whether the military officer corps, the clergy, the law, or medicine. It involved a combination of specialized study, exams, and patronage. If successful, it promised the possibility of ascent from the lowly "gaugers," who counted, weighed, and measured taxable goods, to regional supervisors, excise collectors, and then to the examiners, accountants, and commissioners who oversaw the whole elaborate excise system from the Excise's London offices.

Paine's preparations paid off, and he was awarded a post as a gauger of brewers' casks in the Lincolnshire town of Grantham. Because there was little uniformity in the size of containers brewers used to transport their beer, the gauger had to employ complex mathematics to calculate each container's contents. Paine mastered these techniques and was soon promoted to a better paying position at Alford, a Lincolnshire town closer to the North Sea coast. But Paine's new career turned out to be little better than his earlier one. In 1765 he was discharged from the Excise for "stamping" (this term applied to a range of infractions, usually involving falsified records). Although the offense was a serious one, it was common and rarely clear-cut. Given their poor pay, long hours, and often difficult working conditions, and given that gaugers came to know their clients' practices, the temptation to introduce shortcuts was hard to resist. There was little reason to measure the casks of a brewer whose output rarely changed.

The clearest indication of the ordinariness of Paine's case was his reappointment to the Excise Service in early 1768, this time in the East Sussex town of Lewes. Although he approached his new post with admirable dedication, the burdens of the Excise were overwhelming. In his new posting, the usual ones of low pay and long hours combined with physical danger. The south of England had long been known for its violent smugglers, gangs of whom occasionally turned against the government's tax agents.

There is no evidence that Paine ever faced a direct threat from smugglers, but the work was clearly trying. When some of his fellow officers called on him to help improve the Excise's working conditions, he happily accepted. Paine's primary contribution to the campaign was *The Case of the Officers of the Excise*, his first political pamphlet.

Paine's purpose in *The Officers of the Excise* had very little to do with any philosophical hostility to the British government or its powers of taxation, central themes of his later writings. It was instead to alter the public's perception of the ubiquitous tax agent. Describing that perception four years later in *The Wealth of Nations*, the Scottish moral philosopher Adam Smith would write that tax officers "are naturally unpopular, even where they are neither insolent nor corrupt." The problem was not so much the officers' intrusions into the affairs of individual Britons; it was that the officers seemed to be drawn from humanity's dregs.[2]

This perception, Paine acknowledged, was not entirely unwarranted. Excise officers tended to be a miserable lot, but this had less to do with who joined the Excise Service than with its penurious wages—since "poverty and opportunity corrupt many an honest man." With his pay amounting to fifty pounds annually, ten pounds less than counterparts at the far less onerous Customs, the excise officer was then forced to bear the costs of horse keeping, taxes, rent, ordinary living expenses, and what were known as "removes." Rather than permit too cozy a relationship between gaugers and their clients, the Excise periodically removed gaugers to new and distant jurisdictions. This practice, intended to control corruption, was particularly cruel since it detached officers from the network of friends, family, and creditors to whom even the most ordinary laborer could turn in times of need. "Most poor mechanics, or even common laborers," Paine reminded his readers, "have some relations or friends, who, either out of benevolence or pride, keep their children from nakedness, supply them occasionally with perhaps half a hog, a load of wood, a caldron of coals, or something or other which abates the severity of their distress."

But the excise officer enjoyed no such relations. Just as his family grew acquainted with the people of one town, he was removed to another. Making matters worse, as an officer of the government revenue service, the exciseman faced a further constraint. Any call for financial help could be seen as inviting conflict of interest. There was a fine line between charity and bribery for those charged with overseeing His Majesty's revenues. The common gauger and his family were thus left to endure their desperate circumstances in stoic silence. That is why common perception failed to grasp the true condition of the excisemen. It is a matter, Paine observed, of "reputable pride" and yet a further measure of the sacrifices these officials make on behalf of their employer. As "the thinking part of mankind well knows . . . none suffers so much as they who endeavor to conceal their necessities."[3]

In the end, *The Officers of the Excise* achieved little. It also cost Paine his job. The Excise sacked him for abandoning his post to make the officers' case.

ONCE AGAIN out of work, Paine and his second wife, Elizabeth, the daughter of Samuel Ollive, Paine's former landlord in Lewes, were forced to divest themselves of their possessions. The two eventually separated. Elizabeth found some refuge in the home of her brother, a watchmaker in Kent, and made her way as a dressmaker. By early June of 1774, Paine had left Lewes for London, now with no career, a failed marriage, and few prospects. Sometime during these months of personal crisis, though, his life took a propitious turn. He met Benjamin Franklin.

OF ALL THE AMERICANS the failed stay-maker and former excise officer might have encountered, none could do more for Paine than Benjamin Franklin. Though one of America's richest men, now residing in London as the principal agent for the colonies of Georgia, New Jersey, Massachusetts, and Pennsylvania, Franklin's origins were not unlike Paine's. He had been the child of a Boston soap- and candle-maker and had begun making his way in the world through a menial trade. In

Franklin's case, the trade was printing. Also like Paine, Franklin would work his way out of his trade with literary talent— particularly evident in the clever musings and entertaining doggerel published in his almanacs and newspapers. Finally, Franklin, like Paine, had worked for the British government, albeit in a much more lucrative post, as one of two deputy postmasters for the American colonies.

By the time Paine met him, the middle-aged Franklin had retired from business to a life of public service. It was in this capacity that he worked on behalf of his fellow colonists as they struggled with an ever-escalating controversy over the right of Britain's Parliament to tax its American colonies.

Franklin was no toiling political drudge. He relished London's energy and social vigor. While he spent much time at his Craven Street home, in the company of his landlady, Margaret Stevenson, and her daughter Polly, Franklin also reveled in London's learned society life. "For my own Part," he wrote, "I find I love Company, Chat, a Laugh, a Glass, and even a Song, as well as ever; and at the same Time relish . . . the grave Observations and wise Sentences of old Men's Conversation." He counted among his friends and acquaintances many of London's literary and scientific elite, including Samuel Johnson's printer, William Strahan, and the prominent doctors John Fothergill and John Pringle, the latter of whom would become chief physician to King George III.[4]

When not in his parlor declaiming on electricity and the practical virtues of the lightening rod, Franklin could be found at some of London's famous private clubs, including the Royal Society, England's foremost scientific and philosophical organization. But Franklin's favorite association was a group he called the "Honest Whigs," who met every other week at a coffeehouse near St. Paul's Cathedral. The club was primarily a philosophical society devoted to good cheer and scientific debate. It may well have been at one of its meetings that Franklin first met Paine. Through his many financial ups and downs, Paine had accumulated a network of friends in London that included the Scottish astronomer and popular lecturer James Ferguson. Ferguson was a neighbor and friend of Franklin and very likely introduced Paine to Franklin.

Paine must have seen in Franklin, more than thirty years his senior, an example of what life could hold for those fortunate enough to grasp its fruits. He also must have seen that, for Franklin at least, those fruits were first sown in a land very different from his own. Pennsylvania, as Franklin would explain in his famous *Autobiography*, a book he had begun writing while in England, was the ideal place for a man of keen mind and high ambition. It abounded in fertile farmland, dense, productive forests, hills filled with valuable minerals and game, and a population of hardworking, pious Protestants. It was also home to the world's most modern city.

Unlike London, Philadelphia had no teeming slums and few drunken mobs; its streets were laid out in an orderly grid and they were clean and uncongested. For a man of Paine's inclinations, perhaps the city's greatest virtue was what it lacked. In London, Paine's talents were, simply put, commonplace. The city swarmed with ambitious and bright young people, competing for recognition in business, publishing, and the arts. In Philadelphia, the situation was quite different. There, at the western edge of the British empire, immigrants poured into the city. But most of them were transient. They arrived, perhaps worked for a short time, and then took their talents to the countryside, where cheap American land promised a level of financial security impossible in the Old World. For those with talent but without the taste for country life, Philadelphia was thus an ideal destination. A literate man could find employment, perhaps as a schoolmaster or clerk. And much as Franklin had done, a person could parlay such employment into more fruitful ventures, perhaps in the burgeoning printing and publishing trades.

However much Franklin's life reflected the virtues of his home city and however much those virtues may have inspired Paine to follow a similar path, the fact remains that Franklin had come to love London. By 1774, the year he and Paine met, he had been there continuously since 1757, aside from a two-year trip back to America. He may well have spent the rest of his life in England, entirely abandoning his wife and daughter, who for all those years had remained in Philadelphia. But things in London had begun to go wrong for Franklin.

For the past decade, as Americans rejected one act of Parliament after another, Franklin's wit and charm, his uncanny gift for statecraft, and his general devotion to the British empire all earned him the ear of prominent officials. For years, he had been telling them that minor adjustments of policy would appease the impossible Americans. But on the evening of December 16, 1773, in a catastrophically timed act of resistance, a group of Bostonians raided three privately owned ships, and threw their cargo—342 chests of East India Company tea—into the harbor. This assault on the private property of a British trading company, in whose financial fate the government was deeply invested, suggested to Prime Minister Frederick North that nothing would placate the colonists. In the city of Boston, at least, they had been seduced by the most radical, fringe political elements, people who clearly had no regard for the rights of property or the authority of government.

The timing of the Boston Tea Party could hardly have been worse for Franklin. To the British government, it suggested that all his reassurances about the colonists' fundamentally good intentions were so much window dressing. In other circumstances, Franklin might have been able to rebuild trust—he had done so many times before. But almost precisely coincident with the tea crisis, Franklin would begin fighting to defend himself from accusations of treachery. This was the result of a terrible miscalculation on his part.

Struggling to reassure his colonial brethren that the British government was not conspiring to deny them their liberties, Franklin circulated some purloined letters that included the correspondence of a former governor and lieutenant governor of Massachusetts. The letters had been written between 1767 and 1769 and they contained inflammatory passages calling for the punitive expansion of British government power in the colony. Franklin hoped the documents would convince the colonists that their hostility to Parliament was misplaced and that the sources of Anglo-American tensions were a few rogue colonial officials, some of whom were themselves colonists.

But instead of placating the colonists, the letters inflamed an already deep sense of betrayal; and Franklin's associates in the British government, in particular Secretary for North American Affairs Lord Dartmouth, saw his actions as making a bad situation worse.

In late January 1774, Franklin was called before the King's

advisors on the Privy Council. He was there ostensibly to discuss a petition submitted by the Massachusetts Assembly requesting the recall of Governor Thomas Hutchinson, author of some of the letters in Franklin's bundle. Just before Franklin was to appear, news of the Boston Tea Party reached England. Capitalizing on the ensuing outrage, the Council turned the hearing into a moment of spectacular political theater. Before a restless crowd of government officials and interested spectators, the King's solicitor general, Alexander Wedderburn, savagely attacked Franklin, channeling the government's full fury at the most famous American in the room. Franklin sat in stunned silence.

When Paine met Franklin several months later, the two thus had more in common than modest beginnings and literary talent. They were both entangled in life-changing crises: for Paine, unemployment, a failed marriage, and financial disarray; for Franklin, the ruin of one of his life's projects—the salvation of the British empire in North America. As Franklin contemplated his political collapse, his thoughts naturally turned to his home in Philadelphia, and when Paine consulted the great man about his own future, Franklin suggested that he begin life over again. There was no better place for this former tax agent than Franklin's home city.

HUDDLED ALONG the western shore of the Delaware River, Philadelphia's wharves and storehouses faced a waterfront crowded with sailing ships. The city's port was the busiest in North America. In its markets and shops could be found goods from around the Atlantic, including textiles, housewares, wine, salted meats, manufactured tools and farm implements, teas, spices, and human beings, sold as slaves and servants. "Already it is really possible to obtain all the things one can get in Europe," wrote one visitor. "Ships come in from Holland, Old and New England, Scotland, Ireland, Spain, Portugal, Maryland, New York, Carolina, and the West and East Indies." Facing away from the water, along Philadelphia's streets, one had a view "of the woods beyond the Schuylkill . . . and the woods on the Jersy [sic] Shore." It was as if the city had been carved from the wilderness, and now survived on the lucre of seaborne trade.[5]

In fact, however, that wilderness was an illusion. The woods surrounding Philadelphia concealed the actual source of the city's commercial vibrancy. By 1774, North America's richest and most productive grain-producing region had come to depend on Philadelphia's commercial markets. The western reaches of Pennsylvania, now stretching beyond the Susquehanna River to frontier counties like York and Cumberland; the northern heights of the Delaware and Schuylkill River valleys; much of southern and central New Jersey; much of Delaware and northern Maryland—all channeled their commerce through Philadelphia. Many of the farm families who populated these regions had passed through the port of Philadelphia in transit from Europe; the grain and flour they produced, along with timber, beef, ginseng, iron ore, and other goods, left North America through the port of Philadelphia. And the clothing they wore, the books and newspapers they read, the hoes and axes with which they cleared their lands, were all either manufactured in or imported through Philadelphia.

No other North American city drew on so large a commercial basin. The city's hinterland encompassed nearly 20,000 square miles, or about one-fifth of the entire landmass of the thirteen British North American colonies. The region's population was roughly one-sixth of mainland British North America, or 375,000 people. For the moment, New York City, the nearest major port, could not begin to rival Philadelphia. For its maritime trade, New York competed with towns and cities to the north and east, from New Haven to Boston and Salem; for its inland trade, it competed with those same towns, but also with Albany, Montreal, and of course Philadelphia. But Philadelphia's merchants had a near monopoly on the trade of colonial North America's great middle.[6]

To Philadelphia's west, in the booming agricultural counties of Chester, Lancaster, York, and Cumberland, Germans and Scotch-Irish from Protestant Ulster had created the breadbasket of the Western Hemisphere. Along with farmers in Delaware, northern and eastern Maryland, and southern New Jersey, they transformed Philadelphia into America's first supermarket to the world. In 1772, more than half of all flour exported from the mainland American colonies came through Philadelphia. The region's exports made their way from the port of Philadelphia to other colonial ports and across the Atlantic to Great Britain and ports around Europe and the Mediterranean.[7]

As Paine would discover, this prosperity had transformed a small provincial town into a thriving urban center. With a population of just over 30,000, Philadelphia was the largest city in British North America. It was also North America's most civic-minded city. Through the eighteenth century, Philadelphia's prosperous artisans and merchants founded a host

of civic institutions, including lending libraries, a philosophical and scientific society, a college, and British America's first
medical school and hospital. In a unique expression of volun-
tarism and benevolence, Philadelphia's citizens came together
to form seventeen volunteer fire companies. They also formed
a fire insurance company to protect property from a scourge
they knew all too well, and they established the world's first
antislavery society, in 1775.

Reflecting its cosmopolitan character, Philadelphia had also
become the capital of the American printing trade. By the mid-
1770s, more books were printed there than in any other American city, and by century's end Philadelphia had risen to the
ranks of Dublin and Edinburgh as a publishing center. The only
city in the English-speaking world to surpass this trio for book
publishing was London. The extraordinary growth of the book
business in Philadelphia fueled a new domestic literary market,
and by the 1790s, the number of first editions printed in the city
was almost double the total for Boston and New York.[8]

Along with its many taverns, the city's bookstores and print
shops had become social hubs. Paine found his way to one such
shop, run by Robert Aitken, a Scottish immigrant and former
bookseller in Paisley who had come to Philadelphia several
years earlier. Seizing on the opportunity to bring Americans
the popular religious and philosophical writings of the Scottish enlightenment, whose luminaries included the philosopher
David Hume, the social anthropologist Adam Ferguson, and
Adam Smith, Aitken established Philadelphia's largest book-
shop. Although much of his stock was imported from Scotland and England, Aitken also published works by well-known
Americans, including fellow Scot John Witherspoon, the pres-

ident of the College of New Jersey (later to become Princeton). In conversation with Aitken and his lettered clientele, Paine came to know the American book trade.[9]

Above all, he learned that few eighteenth-century Philadelphians could afford books in large numbers. The city's libraries provided a more affordable way to access books that normally had to be purchased with hefty prepublication subscriptions. Paine also learned that the gentlemen's magazine was a lower-cost alternative to books and, for this reason, promised new opportunity for Philadelphia's printers.

Much lengthier than newspapers, which rarely contained more than short news stories, shipping news, and advertisements, these periodicals offered a bounty of practical and moral knowledge. Their subjects could range from viticulture to philology, and from the morality of slavery to the philosophy of education. For the most part, gentlemen's magazines avoided open religious or political partisanship, preferring to respect the kind of ecumenical republic-of-letters ethos that in the mid-eighteenth century governed conversation in coffeehouses and learned societies. Nearly all such magazines in the colonies came from Britain. Philadelphia printers had attempted to launch four beginning in 1741, but none lasted much more than a year.

By the fall of 1774, Aitken was beginning to see bright prospects for a new American magazine. The nearly decade-long clash between America and the British Parliament over taxes had generated a series of boycotts in the colonies of paper goods, teas, textiles, and assorted other manufactured imports. Perhaps, Aitken reasoned, all the uncertainty in trade would allow him to underprice British magazines. The thinking was commonplace in Paine's Philadelphia and accounted for a remarkable surge in American manufactures.

Though it had gone out of business by the time Paine arrived, the Bonnin & Morris ceramics manufactory exemplified this trend. Its porcelain tableware, equal to the finest European imports, allowed Philadelphians to avoid the tangled politics of importing ceramics from Britain. Not far from the city, in Lancaster County, entrepreneurs created one of America's first glassworks. And in what was among the boldest, most self-conscious efforts to promote American manufacturing, in 1775, a group of merchants, mechanics, and artisans banded together to found a textile company called the United Company of Philadelphia for Promoting American Manufactures. The business would be among Philadelphia's largest and most profitable, and would stand as a direct rebuke to Britain's own textile manufacturers. Americans, it seemed, could do for themselves what Britain had long done for them.[10]

Drawing on this same spirit of independence, Aitken decided to publish a new American gentlemen's magazine. The time was right. But his other business concerns meant that he could not be responsible for his magazine's content. For this, he would need a full-time editor, and in Thomas Paine he found precisely the man for the job. Although Paine had published only a few newspaper articles and a single pamphlet by this time, he had other qualities that made him suitable. He knew Aitken's intended audience. His growing Philadelphia circle and his acquaintances in London and Lewes were all the kinds of ambitious, aspiring artisans, mechanics, merchants, and minor gentlemen such a magazine was designed to attract.

DURING ITS NINETEEN-MONTH LIFE, from January 1775 through July 1776, Aitken's *Pennsylvania Magazine* was the only substantial periodical published in Philadelphia. In London, there were more than thirty such magazines, absorbing the scribblings of countless writers. These ranged from the aristocratic Lady Mary Wortley Montagu to the Irish poet and novelist Oliver Goldsmith, who began his literary career as an impoverished hack writing reviews for the London literary magazine *The Monthly Review*.[11] The sheer number of such periodicals was a testament to reader demand and the abundant supply of ever-churning Grub Street writers. Though Philadelphia did have its own provincial literary scene, those with the talent and wit to see their prose into print could find easier paths to security and prosperity, particularly in the law, which was undergoing a revolutionary expansion of its own. But for a man in his late thirties who had already tried several careers and whose financial circumstances were precarious, the magazine business was about as good as could be hoped for. Indeed, it turned out even better than that. The clear language, the well-chosen simile or metaphor, and the biting sarcasm that made Paine's political writings so popular served him well in his new work.

Within months of starting at *The Pennsylvania Magazine*, Paine could report to his friend Franklin that Aitken "had not above six hundred subscribers when I first assisted him. We have now upwards of fifteen hundred, and daily increasing." Paine had found his calling. With few others prepared to devote themselves to a learned periodical, Paine was forced to write much of the content in *The Pennsylvania Magazine*—exactly how much is impossible to say, since so much was pseudonymous.

And Paine had a hand in selecting whatever he did not write. It was Paine the editor who was responsible for including essays such as "A Description . . . of a New Invented Spinning Machine" or "The Effect of Musk, in Curing the Gout in the Stomach."[12]

An introductory essay in the magazine's first issue, surely written by Paine, presents the new publication to its readers as essential in a "country whose reigning character is the love of science" but "whose channels of communication should continue so narrow and limited." The magazine would do much to widen those channels, and there was no better moment for doing so. "America has now outgrown the state of infancy," and was ready to sustain a learned magazine:

> The failure of former [magazines] cannot be drawn as a parallel now. Change of times adds propriety to new measures. . . . A magazine can never want matter in America, if the inhabitants will do justice to their own abilities. Agriculture and manufactures owe much of their improvement in England to hints first thrown out in some of their magazines. Gentlemen whose abilities enabled them to make experiments, frequently chose that method of communication, on account of its convenience. And why should not the same spirit operate in America? I have no doubt but of seeing, in a little time, an American magazine full of more useful matter, than I ever saw an English one: Because we are not exceeded in abilities, have a more extensive field for enquiry: And whatever may be our political state, OUR HAPPINESS WILL ALWAYS DEPEND UPON OURSELVES.

As the American struggle with Parliament continued and as Philadelphians confronted the possibility of war and revolution, these words, from a man in their country but a few months, must have been inspiring. The inventiveness from which their English brethren drew so much economic power had simply lain dormant in their own country, but American ingenuity was beginning to stir, and when it finally emerged, would bring about a revolution in agriculture and manufacturing.[13]

In pursuit of this object, *The Pennsylvania Magazine* availed readers with tales of a method for the "Stilling of Waves by Means of Oil," of "An Easy Method to Prevent the Increase of Bugs," and of "A Method of Making Mortar Which Will Be Impenetrable to Moisture."[14] In addition to useful knowledge, readers found extensive discussions of popular morality—discussions about marriage, education, and the proper order of society. While they found little direct political commentary, they did find much that was clearly political. And most of this bore the marks of Paine. "Reflections on Titles," for example, attacked one of the Old World's most ingrained habits:

> The title of *My Lord*, over-awe[s] the superstitious vulgar, and forbid[s] them to enquire into the character of the possessor: Nay more, they are, as it were, bewitched to admire in the great, the vices they would honestly condemn in themselves. This sacrifice of common sense is the certain badge which distinguishes slavery from freedom; for when men yield up the privilege of thinking, the last shadow of liberty quits the horizon.[15]

The essay never openly links the absurdity of aristocratic titles to the ongoing American struggle with Britain. But the underlying message could not have been clearer: societies that accepted the artificial distinctions of title were societies that accepted their own enslavement. Societies that rejected them were free. Americans would soon be forced to choose between the two.

2

The Hazards of Competition

PAINE ARRIVED IN PHILADELPHIA at its colonial apex. In coming decades, as the American population pushed farther west, the advantages afforded by Philadelphia's vast hinterlands began to decline. Much of the prime farmland adjoining the Delaware, Schuylkill, and Susquehanna Rivers had been claimed, and migrants were making their way to the Allegheny and Ohio River valleys. Meanwhile, the colonial commerce that had moved from the hinterlands of the middle colonies to the Philadelphia waterfront was beginning to flow to another, newer city.

Well before Paine's arrival, Philadelphians had recognized that however advantageous the city's geographic circumstances, however inventive its artisans and merchants, and however progressive its civic leaders, the city's commercial primacy was uncertain. Farmers, fur traders, and ironworkers along the Susquehanna River had begun taking their produce to a fledgling commercial town to the south. In 1759, a Swedish visitor to Lancaster County observed that the pig iron produced in the region's foundries was not sold in Philadelphia's markets, but was "carried to the Susquehanna River, thence to Maryland, and finally to England."[1] By the early 1770s, this commerce transformed Baltimore Town from a tiny hamlet, smaller even than Annapolis, into the largest town in Maryland. In 1770 its population was still only 6000, just more than half that of Charleston and about two-thirds that of Newport. But its boom had begun. By 1800, Baltimore would be America's fastest growing city. Ten years later, it would be the third largest city in the country with a population of 46,000; only Philadelphia and New York were larger.[2]

The reason for Baltimore's early growth was transportation. As Pennsylvanians had long known, there was no easy path from the Susquehanna and its basin to Philadelphia. Backcountry producers on the river's east side could ship goods overland. But as one observer wrote, Pennsylvania's roads were often "rendered almost impassible by the multitude of carriages which use it." Farmers on the western side of the Susquehanna faced the added challenge of a dangerous and costly crossing of the region's largest river. In the face of such barriers, it was no wonder that "Baltimore town, in Maryland, has within a few years past carried off from [Philadelphia], almost the whole of the trade of Frederick [in Maryland], York, Bedford, and Cumberland counties," all west of the Susquehanna. For western Pennsylvanians, in other words, it was simply cheaper and faster to move goods south, along the Susquehanna to Chesapeake Bay and the new port of Baltimore.

In 1771, four years before the Revolutionary War erupted outside of Boston, Samuel Rhoads, a Pennsylvania assemblyman and home builder, explained the problem to Franklin, his most famous client: "The growing *Trade* of Baltimore Town in Maryland drawn principally from our Province west of Susquehanna begins to alarm us with serious Apprehension of such a Rival as may reduce us to the Situation of Burlington [New Jersey] or New Castle on Delaware." To avoid becoming a second-tier town, Philadelphia would have to draw westerners to its markets. A canal linking the Susquehanna and Delaware Rivers was the most obvious way to do this. Whether or not such a colossal undertaking was at all possible was another matter. For some time, Philadelphia's merchants had been exploring the idea and in 1769 they put the canal question to colonial America's foremost scientific society.[3]

The American Philosophical Society, begun in the early 1740s by Franklin, the Pennsylvania farmer and botanist John Bartram, and the physician Thomas Bond, found new life in the decade before the Revolutionary War. As relations with Britain deteriorated, and Americans began confronting life without British manufactured goods, the society's original mission of "Promoting Useful Knowledge" acquired a new urgency. Since many of the society's leading members were Philadelphians, closely connected to the city's merchant community, it took a leading role in the canal project. Here was science in the service of Philadelphia's commercial empire.

In its preliminary studies, the society concluded that a new artificial water route would require the excavation of several hundred thousand cubic yards of dirt at a cost of between $900,000 and $3.6 million (in today's dollars). The variation in cost reflected uncertainty over whether the canal ought to accommodate small, flat-bottomed barges, or deeper-keeled sailing vessels, capable of carrying more goods, and doing so in the open waters of the Chesapeake and Delaware Bays. The latter would require the construction of stone locks and a canal deep enough to accommodate the ships' deep drafts. Barges could simply be hauled from shore up and down the shoals and rapids that would inevitably form in a shallow canal.

But barges carried additional costs, unrelated to the construction of the canal. For goods to be moved from the canal's mouth out into Delaware Bay, and up the river to Philadelphia, they would have to be transferred to more stable, seaworthy craft. Warehouses and wharves would have to be built to accommodate this transfer of goods. Low-wage stevedores would have to carry the goods from barge to ship. With many of those very same workers demanding better wages and greater political participation in exchange for their cooperation with colonial anti-importation agreements, militia musters, and other acts of resistance to British authority, the gentlemen of the Philosophical Society surely reckoned this a high cost.

And then there was the Susquehanna itself. The farther south one traveled, the more difficult it was to navigate, especially in the drier summer and fall months, when the Bald Friar Falls just south of the Pennsylvania border emerged out of the river's rocky bottom. Any of the southerly locations of the canal would involve the additional construction of a safe watercourse through these falls. Another site that seemed to accommodate a relatively short and lower-cost canal was found likely to "carry all the navigation of the river Susquehanna . . . [too] far down the Chesapeake Bay, for an advantageous communication with Philadelphia."

These problems were sufficiently vexing that some Philadelphians began exploring alternative canal routes. To Samuel

Rhoads, the solution to the Baltimore problem was a northern route, linking the "Susquehanna to Schuylkill . . . so far as they lead towards our Capital City" of Philadelphia. These would all have been enormous and costly undertakings. In the face of economic disruption caused by the ongoing political struggle with the mother country, there is little reason to think that Philadelphians could have made them a reality. But the elaborate exploration and planning suggests a business community firmly convinced that its commercial well-being depended on ambitious internal improvement.[4]

BY THE TIME Paine arrived in the fall of 1774, Philadelphia's long-term commercial prospects had drifted from the minds of its citizens. Now the preoccupation was more immediate. A Parliament beholden to special interests and a monarch oblivious to the deceptions of his ministers appeared bent on transforming colonists from tolerant, liberty-loving Englishmen into beleaguered servants of a grasping, greedy empire. The burning question now was not whether Philadelphia would continue to prosper, but whether it would continue to be part of imperial Britain.

For their brethren in Massachusetts, the eruption of open warfare on Lexington Green in April of 1775 would answer that question. Bostonians, who had already endured a lengthy British military occupation, had little doubt that Britain would use any means to subdue its colonies. For many, Lexington confirmed that independence was now the only viable path. There was no reconciling with a king prepared to turn his army against his own subjects.

In Philadelphia, sympathy for the Bostonians was widespread, but the march to independence was slower. There had been no occupying British army to unify public opinion against the King, and patriot voices struggled to be heard over the din of long-standing political divisions. The city's merchant elite often clashed with its artisans and shopkeepers over issues of trade and taxation. Moderate Quakers and Anglicans had long fought with evangelical Presbyterians; English-speaking colonists questioned the loyalties of Pennsylvania's German speakers. Across all these fissures lay the long simmering conflict between supporters of the colony's proprietors, the Penn family, and their opponents, led by Benjamin Franklin.

After war erupted near Boston, the tide began to turn. Franklin, just returned from England, gave the more radical anti-British partisans a powerful new ally. Meanwhile, the British did their part to encourage colonial resistance. The burning of Charlestown, Massachusetts, and the bloody Battle of Bunker Hill in mid-June of 1775 were followed by a series of failed colonial efforts at reconciliation. The best known of these was the so-called Olive Branch Petition initiated by John Dickinson, a member of Pennsylvania's delegation to the Continental Congress, the ad hoc body that colonists had initially created to coordinate their response to British taxation but which was now overseeing the war effort. Dickinson was also a leader of the moderate wing of the patriot cause. He and his allies hoped the petition would persuade the King to end the war and negotiate a political settlement with the Americans. But the gesture only produced more British truculence.

On August 23, 1775, George III formally proclaimed the American colonies in open rebellion. Any of his subjects found to be aiding the American rebels would now be tried for treason. The news took several months to reach the colonies, but when it did in early November, it confirmed what radicals had long believed. The problem with Britain was not confined to a group of rogue ministers and a Parliament run amok. The problem was with the monarchy itself. Moderation, it seemed, was heading for extinction. The King's proclamation "has a most happy effect," observed Samuel Ward, a Rhode Island delegate to the Continental Congress, because "those who hoped for Redress from our Petitions now give them up & heartily join with us in carrying on the War vigorously."[5]

FOR MOST OF HIS first year in America, Paine wrote very little about the revolutionary crisis. But as events drove more Philadelphians to the side of resistance and even independence, Paine began to speak out. And he began to say something many had come to believe, but few dared write. The only reasonable way to preserve American rights and liberties was independence.

In October 1775, just before word arrived of King George's incendiary proclamation, in a brief newspaper essay innocuously entitled "A Serious Thought," Paine offered his first public pronouncement on American independence. In the resonant, clear-eyed language for which he would become well-known, he reminded his readers that their recent problems were but the tip of an iceberg—an iceberg whose subsurface cruelties stretched from America to India. Under the guise of the quasi-private British East India Company, Britain's India policies produced famine and bloodshed, the latter in the form of murderous rampages against Indians who refused to fight the company's battles. In Africa, meanwhile, the company's merchants carried on a barbaric trade whose destructive consequences were inescapable, even for those in far-off America. Britain's evils were so extensive and insidious, there was but one path for right-thinking Americans:

> When I reflect on the horrid cruelties exercised by Britain in the East Indies—How many thousands perished by artificial famine. . . . When I read of the wretched natives being blown away, for no other crime than because . . . they refused to fight—When I reflect on this and a thousand instances of similar barbarity. . . . —And when to these and many other melancholy reflections I add this sad remark, that ever since the discovery of America she has employed herself in the most horrid of all traffics, that of human flesh unknown to the most savage nations, has yearly (without provocation and in cold blood) ravaged the hapless shores of Africa, robbing it of its unoffending inhabitants to cultivate her stolen dominions in the West—When I reflect on these, I hesitate not for a moment to believe that the Almighty will finally separate America from Britain. Call it independence or what you will, if it is the cause of God and humanity it will go on.[6]

As long as they remained subjects of the King of England, no Philadelphian would be spared the taint of Britain and its empire.

IN LATE 1775, AS the December freeze descended and the costs of months of warfare piled up, many of Paine's fellow Philadelphians remained unable or unwilling to grasp this logic. For some, the problem was the old one of economics. Without British trade, American business would suffer. But for others, the concerns were more far-reaching. To separate from the United Kingdom was to challenge the political wisdom of centuries. Many of the most astute political minds of the day considered hereditary monarchy the only way to political stability. A country without a king was like a roof without walls: it was doomed to collapse. Lacking the unity imposed by the British crown, the American people might quarrel among themselves and descend into civil war. Lacking royal naval or military protection, they would be vulnerable to Britain's French and Spanish rivals, never mind the great military might of Britain itself. Former colonies lacking a true head of state would also be unlikely to gain membership in an international community of monarchical nations.

Such fears were widespread in the Pennsylvania General Assembly, which ordered the colony's Continental Congress delegation to oppose any move formally to sever ties with Britain. Meanwhile, Pennsylvania's delegation continued to support the war effort. For the ever-diminishing number of moderates, there was no contradiction here. They hoped that the burdensome costs of war would soon lead the King and Parliament to the negotiating table. Even after the rejection of the Olive Branch Petition, and subsequent indications that the King was preparing to expand the military campaign, moderates maintained their opposition to independence.

For Paine and other radicals, including the famous Massachusetts delegates to the Continental Congress, Samuel Adams and his young cousin John, the persistence of moderation was inexplicable and infuriating. They saw no sign that the British government would reconcile on terms other than its own. Meanwhile, as the war continued, Congress was forced to consider some kind of military alliance, most likely with France, the only power whose military resources could begin to rival Britain's. But the French were reluctant to support the Americans as long as reconciliation with Britain remained a possibility. It all made for a classic political stalemate—radicals saw independence as vital to military success while the moderates saw in independence a road to ruin. Further complicating matters, as it became a war-making body, in November 1775, the Continental Congress imposed new secrecy rules. The most influential politicians in the colonies were now forced to adopt a stately reticence. The business of Congress could no longer figure in the debate over independence.

Proponents of independence had to find a new way to bring their case to the public. This need stirred Philadelphia's leading radical voice, Benjamin Rush, a physician, professor of chemistry, and antislavery activist, to enlist Paine, whom Rush had met through Aitken, and who had shown himself ideally suited to make that case. "I perceived with pleasure," Rush recalled of their conversations, that Paine "had realized the independence of the American Colonies upon Great Britain" was now the only certain way "to bring the war to a speedy and successful issue." Rush remembered suggesting to Paine that he make the case in print. The doctor would have done so himself, but he "shuddered at the prospect of its not being well received." As an evangelical Presbyterian, Rush was unlikely to sway old-guard Anglicans, let alone Philadelphia's many pacifist Quakers.

Rush also had a career to protect. His work as professor and physician demanded a reputation unsullied by political controversy. The Reverend William Smith, president of the College of Philadelphia, where Rush was professor of chemistry, had been a leader among the moderates, and the "great majority" of Philadelphians, some of whom were Rush's patients, "were hostile to a separation of our country from Great Britain." Paine, in contrast, had no reputation to protect. He was new to the country, had no family in America, and was unknown outside of Philadelphia. As Rush put it, "he could live anywhere."[7]

In late 1775, Paine thus composed a bold call for American independence entitled *Plain Truth*. At Rush's urging, he changed the title to *Common Sense*. The pamphlet contained no arcane legal argument; no dry political theory. In the simplest of terms, often drawing on the Bible, the one text all its readers knew, *Common Sense* justified the independence Paine and his allies had come to see as inevitable. Much of that justification came as a searing critique of the British mode of governance, particularly its hereditary monarch. The idea that birth alone entitled a human being to rule was, Paine proclaimed, among history's greatest humbugs. "One of the strongest natural proofs of the folly of hereditary right in kings," he wrote, "is that nature disproves it, otherwise she would not so frequently turn it into ridicule, by giving mankind an *ass for a lion*." The routine incompetence of the likes of George III was proof enough that no natural order would ever sanction governance as a right of birth.[8]

As he turned his attention from magazine editing to politics, Paine's relationship with Aitken deteriorated, and when the time came to find a publisher for *Common Sense*, he looked elsewhere. Robert Bell, another Scottish printer and bookseller, had a shop on Third Street next to the Anglican St. Paul's Church and was a prominent supporter of independence. Bell had also published a number of the enlightenment-era tracts that informed the independence movement. Among these was the conservative English jurist William Blackstone's *Commentaries on the Laws of England*, which contained a defense of the people's right to replace a monarch who acts against English constitutional principles. Bell agreed to publish *Common Sense* at his own expense and on January 10, 1776, the first copies began to issue from his press.

Nothing Bell—or any other colonial printer—had ever published approached the impact of *Common Sense*. Indeed, nothing anybody in the English-speaking world had ever published had a publishing history quite like *Common Sense*. Within a year, tens of thousands of copies were in circulation—many times that of an ordinary pamphlet. In 1776 alone, *Common Sense* went through nearly forty printings in the colonies and about half that number in Britain. The pamphlet's popularity was a measure of its success. "Its effects," Rush recalled, "were sudden and extensive. . . . It was read by public men, repeated in pubs, spouted in schools and in one instance delivered from the pulpit instead of a sermon." One Philadelphian recalled that "*Common Sense* . . . is read to all ranks; and as many as read, so many became converted; though perhaps the hour before were most violent against the least idea of independence." A New York newspaper noted, "A pamphlet entitled *Common Sense* has converted thousands to Independence, that could not endure the idea before." And several years later, during a diplomatic mission to France, John Adams was stunned to find that "the pamphlet *Common Sense* was received in France and all Europe with rapture."[9]

GIVEN ITS enormous popularity, *Common Sense* could have been a path to financial security for Paine. But at a time when he was asking Americans to make the terrifying journey to independence, there could be no profit amid so much danger. Better to show that, contrary to what most of his critics would contend, his was not an act of opportunism. That "the sun never shone on a cause of greater worth" than American independence was something Paine would have to defend with both his pen and his person.[10]

He donated his share of the profits from *Common Sense* to the desperate Continental Army, established by the Continental Congress after the outbreak of fighting in 1775. He also financed a second cheaper edition with his own meager savings. In an even bolder demonstration of commitment to cause, Paine would join the Continental Army. Paine's Quaker origins, his recent predilection for urban, literary employment, his now familiar rhetorical powers, never mind his age, would all appear to make him among the least likely citizen-soldiers Philadelphia could muster. But just days after the Continental Congress submitted to the world a formal Declaration of Independence, he marched off to war.

3

Years of Peril

G IVEN HIS SKILLS AND HIS AGE, Paine would endure little
of the hardship so familiar to the common soldier. But his
brief time in the field of battle would nonetheless prove difficult. He served first as secretary to General Daniel Roberdeau,
a wealthy merchant from Philadelphia who commanded a
large group of Pennsylvania volunteers. Roberdeau's force was
what was known as a "flying camp," or a group—ten thousand in this case—committed to "fly" to a single battle zone,
then quickly disband at battle's end. During August and early
September 1776, as British forces arrived at Staten Island in
preparation for the invasion of Manhattan, Paine and the rest
of Roberdeau's men looked on from Perth Amboy in New Jersey. Before ever seeing action, Roberdeau's volunteers began
returning to Pennsylvania, their battle having never materialized. Roberdeau was forced to disband the remainder of his
troops, but Paine pushed on. He would join General Nathanael
Greene as aide-de-camp at Fort Lee, on the west side of the
Hudson River. While the British drove Washington's army out

of Manhattan to White Plains, Paine, Greene, and Greene's men could only watch in horror.

By early November, British forces commanded by Major General Charles Cornwallis had surrounded Fort Washington, situated across the Hudson from Fort Lee. Within days, the Fort Washington garrison of twenty-eight hundred had been captured. Greene and his army would have to flee, dodging a surprise British attack and barely surviving a fierce bombardment in New Brunswick, New Jersey.

By the beginning of December, the Continental Army was in full retreat, backtracking helter-skelter through the New Jersey countryside as deserters fled, enlistees returned home, and the ranks of the army dwindled to three thousand badly armed, poorly clad, malnourished souls. Meanwhile, British forces under the command of Cornwallis pursued the Americans, facing little resistance along the way. Fearing entrapment against the eastern bank of the Delaware, Washington ordered the remains of his forces back to the relative safety of the river's western shore. With the army in fast retreat, and the British moving seemingly unmolested toward the new nation's capital at Philadelphia, Paine and the military had begun to fear that not only would the British take the city, but they would also close down the chief instrument of patriot propaganda, the Philadelphia printing trade.

In advance of this final calamitous circumstance, in early December Paine left the army at Trenton and traveled alone and on foot back to his home. It was a brutal winter's journey of some thirty-five miles, and its conclusion was terribly depressing for Paine. Congress had decamped to Baltimore while patriots fled for the countryside. In place of the vigorous commercial capital he had come to know, Paine found Philadelphia largely deserted, save for some dedicated loyalists, convinced they would soon be liberated by a triumphal British army.

Racing against time, Paine immediately set out to fulfill the mission that had brought him back to the city. He would begin publishing a series of thirteen essays entitled *The American Crisis*. The essays, which appeared periodically over the course of the next six years, were as vital to the cause of American independence as *Common Sense*. "A few days after our army had crossed the Delaware on the 8th of December, 1776," he recalled, "I came to Philadelphia on public service, and, seeing the deplorable and melancholy condition the people were in, afraid to speak and almost to think, the public presses stopped, and nothing in circulation but fears and falsehoods, I sat down, and in what I may call a passion of patriotism wrote the first number of *The Crisis*."[1] Despite the many difficulties facing the printing trade, Paine managed to see the essay into print just before Christmas of 1776, first in the pages of the *Pennsylvania Journal* and then as an eight-page pamphlet.

At a time when the American military campaign seemed to be falling apart, Paine's first *Crisis* provided what the patriot cause needed most: it reminded Americans that theirs was the just cause. "Tyranny, like hell, is not easily conquered," Paine famously intoned. But "what we obtain too cheap, we esteem too lightly." Nothing worth the fight would come without a cost. Those inclined to abandon the cause of American liberty in these dark moments simply did not grasp this essential truth.

As had become his habit, Paine eschewed any income from his writings, returning his earnings to the printers, provided they continue printing the pamphlet. This they did, producing an initial print run of 18,000. Paine was now the chief propagandist for the American military campaign. Over the course of the next two years he would publish six more *Crisis* essays, but the first of them could not have come at a more propitious moment.[2]

Paine himself described the winter of 1776 as "the very blackest of times."[3] On one side of the Delaware River stood a depleted American force of about twenty-five hundred men and on the other, a stronger and much better armed British force of around five thousand. Had Washington not kept all available boats and ferries from enemy hands, the situation might have been even more desperate. Now the two armies stared at each other across the broad waters of the Delaware. All the while, the few remaining patriots of Philadelphia trembled, aware that an invading force was one river crossing from their city.

Washington worried that enemy troops would now simply camp out in New Jersey, buying time until a frozen Delaware River could provide a path to the capital. "I do not see what reasonable hope there is to preserve Philadelphia from falling into the Enemy's hands," he wrote Major General William Heath on December 21, 1776. "They will attempt to possess it as soon as the Delaware is frozen as to admit their passage."[4]

The Continental Army had already proven no match for British forces, and with patriots leaving Philadelphia in large numbers, whatever militia the city could muster would be limited. The defense of the capital would fall to Washington's Continentals, who needed time to prepare. As winter descended, time was precisely what Washington lacked. After the first of the year, when the river would begin to freeze, enlistments would begin to expire and his army's numbers would plummet.

Adding to Washington's troubles was the hydrography of

the Delaware. The river tended to freeze over in January, but there was no way to be certain when and where it would freeze. Should that moment come later, a much depleted Continental Army would offer little resistance to a British assault. Should it come earlier, there would be no time to prepare Philadelphia's defenses. After a particularly bad cold spell a few days before Christmas, an early freeze came to seem likely. Rather than risk the nation's capital, Washington chose to act.

On Christmas Eve, he convened a meeting of his advisors to begin planning a preemptive attack on the British camped across the Delaware. It was resolved that at nightfall, several thousand troops would cross the river for a surprise attack against fifteen hundred Hessian mercenaries camped near Trenton. It would be a harrowing crossing. The problems Washington's army faced were among the same ones that would later compel Paine to devise safer ways across America's hazardous waterways.

THE CROSSING began late on Christmas Day during a wicked winter storm. Amid blowing rain, sleet, and snow, Washington's army was to make landfall at three points in New Jersey. Washington and the troops directly under his command would land about ten miles upstream from Trenton. A smaller force of about eight hundred was to land just south of the town, near the Trenton Falls, and a third group of about eighteen hundred soldiers would cross to Burlington, New Jersey, about twelve miles to the south.

The broad-bottomed boats the Continental Army used to

ferry troops and supplies were well suited to the tempestuous waters that flowed through the rivers and harbors of the American northeast. But they were no match for river-borne ice. The problem was particularly acute on the Delaware, a tidal river south of the Trenton Falls. In a matter of a few hours, as the tide came in, hundreds of tons of free-floating ice would be pushed back up river toward the falls, where it would accumulate in mile-long jams, some piled five and six feet above the water's surface. There was no way to navigate these ice floes by boat, and any attempt to cross them on foot meant almost certain death by drowning. Similar problems plagued forces farther down the Delaware.

Captain Charles Willson Peale, who had abandoned a flourishing portrait-painting business to serve, recalled that the Philadelphia militiamen under his command made it across to the New Jersey side only to find "the ice gathering so thick at a considerable distance from the shore that there was no possibility of landing." A few of Peale's men were able to break through the ice and land on the New Jersey side, but their numbers were ultimately too small to advance and they were ordered to return to Pennsylvania. By the time they began the return crossing, furious storm winds had kicked up, and the swirling river and its fast-moving ice threatened to crush the boats. Captain Thomas Rodney, whose men were among the few to make it across, recalled that "the wind blew very hard and there was much rain and sleet, and there was so much floating ice in the River that we had the greatest difficulty to get over again, and some of our men did not get over that night." Only the troops directly under Washington's command, to the north of Trenton, made it across in sufficient numbers to carry the campaign forward.

This small victory was largely owing to the heroics of the army's boatmen—some of whom were experienced local ferrymen, others hailing from as far away as Marblehead, Massachusetts.[5]

Pushing and pulling their oars, and using long river poles, they heaved their unwieldy craft—some as long as sixty feet— and somehow delivered the force responsible for the American victory at Trenton.

Washington's troops captured about a thousand Hessian mercenaries. After several days, the Americans had secured the town. From Trenton, Washington's army pushed farther into New Jersey and defeated British forces at Princeton. The victories were modest but their cumulative effects would ultimately drive the British back to New York, temporarily saving Philadelphia. For most of the following year, Paine dedicated himself to supporting the army and promoting the patriot cause through his *Crisis* essays.

WRITING FOR THE CAUSE would become much more difficult when, in late August 1777, British forces landed in Maryland and began marching north toward Philadelphia. As the enemy approached, Paine fled the city. From that point on, any hope he had of secluding himself from the immediate disruptions of battle was gone. By the end of September, when British General Lord Howe's forces occupied Philadelphia, he left the city for the home of his friends Joseph and Mary Kirkbride, who lived in Bucks County, on the Delaware River. In December, he would visit Valley Forge, the rough winter encampment of Washington's army. "I was there when the army first began to build huts," he wrote Franklin, and "they appeared to me like a family of beavers: every one busy; some carrying logs, others mud, and the rest fastening them together. The whole was raised in a few days, and is a curious collection of buildings in the true rustic order."[6]

Paine left the camp soon after, sparing himself the miseries winter would bring to its inhabitants. Instead, he would spend those cold, dark months in the relative comfort of the Kirkbride home. But these would be trying months. The Continental Army had been driven from New Jersey and patriot forces had been unable to prevent the fall of the American capital.

Just before fleeing Philadelphia, Paine had published a fourth, abbreviated *Crisis* essay. Another full-length one would not appear until the twenty-first of March the following year. There was no way for Paine to spread his message with the printing trade, such as it was during the war, now under British control. Nonetheless, this period of exile would provide Paine with a new perspective on his adopted country. Traversing the countryside surrounding Philadelphia, he became acquainted with its geography and its war-time transport infrastructure. Most of this consisted of roads, heavily rutted by wagon wheels and horse hooves, fords across often swollen or icy creeks and streams, and, weather permitting, ferries and ferrymen. During these months, Paine would also become better acquainted with Philadelphia's other river, which had become the chief boundary between American and British forces.

In early December 1777, before it camped at Valley Forge, Washington's army had been forced to retreat back across the Schuylkill River northwest of Philadelphia. While any winter river crossing was difficult, the soldiers were spared the worst. Military engineers had erected two floating bridges at Sweed's Ford, near present-day Norristown. One, described by Albigence Waldo, the army's surgeon general, consisted of "36 wagons with a bridge of rails between each" and another was composed of linked, wooden rafts.[7]

In his journeys through the Philadelphia hinterland, Paine would cross these bridges; early in his exile, he might also have crossed a new floating bridge that had replaced Philadelphia's Market Street or "Middle" Ferry in 1776. What came to be known as the Middle Ferry Bridge was built by the Pennsylvania Committee of Safety, the civilian council responsible for the state's military affairs. The bridge was needed to ease the movement of provisions and personnel in and out of Philadelphia as it prepared for war. At the time the bridge was built, the Schuylkill was of no particular strategic importance. All this had changed when British troops landed in Maryland and began marching north.

In September 1777, as Philadelphia fell, Washington had ordered his engineers to detach the original floating bridge from its moorings and float it into the Delaware. As worrisome to military planners as the capture of the bridge was, the capture of the chain of boats out of which the bridge had been made was equally troubling. It was one thing for an invading force to haul wagons and artillery overland, but to carry its own boats was a rarity, and General Howe's fifteen thousand troops would need some means of crossing the Schuylkill if they were to take Philadelphia. In addition to the removal of the Middle Ferry Bridge, Washington thus ordered "all the Boats of every kind . . . to be placed so as to be entirely out of the reach of the enemy."[8]

The British ultimately crossed the Schuylkill at Flatland Ford, ten miles upriver from Philadelphia. And when they entered the city on September 26, they did so from the north, on land. But soon after taking the American capital, they began construction of their own floating bridge at Middle Ferry.

What made these bridges so essential was their capacity to carry large numbers of troops, wagons, horses, artillery, and provisions. To ferry ten thousand or more men and their equipment, even in dozens of boats, entailed much greater hazard and took much more time. Washington's retreat across the Delaware in early December 1776, which involved some forty boats and a relatively small force, had taken five days and nights.

Efficient though they were, floating bridges required boats and rafts across which to lay a roadway. For a bridge across a river the size of the Delaware, it was difficult to find enough boats and rafts to stretch the full width of the river, and even if enough could have been found, assembling them all in one place heightened the risk of capture. A floating bridge also had to be anchored in place, lest the roadway's undulations make it impassable. A river that was too wide or too deep made this difficult.[9]

In the winter there was no need for bridges at all. Once the river froze—and as a shallow, relatively modest, and slow-moving river, the Schuylkill did so much earlier and more definitively than did the Delaware—it could be easily traversed. Indeed, the freezing of the Schuylkill presented Washington and his generals with a whole new set of problems. Although there was less forage to be found as winter advanced, a frozen Schuylkill meant that British troops would be readily able to tramp through the countryside in search of grain stores, livestock, and vulnerable patriot forces.

But a frozen river also gave Washington's forces the opportunity to cross into Philadelphia from the west, and at the beginning of December Washington and his generals debated the feasibility of such an assault. The entire discussion centered on the reliability of the Schuylkill's winter ice. At some point, everybody knew, the Schuylkill would freeze, but nobody knew when or for how long. A brief cold spell was almost as bad as no cold spell at all, raising the danger of river rapids and dangerous free-floating ice. The young French nobleman the Marquis de Lafayette, who had joined the American cause in June 1777 and who had quickly earned Washington's fatherly affections, explained to his commander that even if a frozen river would support an invading force, any European army would know a quick and easy defensive measure. "In Europe ice is broken every night when it can facilitate the projects of the enemy," and "if all is not cleared, at least a ditch can be formed in the river." Even if the ice could not be fully broken, a channel could be made to slow wagons and marching soldiers.[10]

Given the uncertainties the river presented, Washington abandoned plans to liberate Philadelphia, choosing instead to wait out the British until the following spring. In the end, the decision was the right one. With France's entry into the war in early 1778, British strategy shifted. The army would retreat to heavily loyalist New York and then begin a new campaign to take the southern colonies. As part of this strategic shift, Philadelphia would be abandoned, and so, by the end of June 1778, the new commander of British forces in North America, General Henry Clinton, led his soldiers, along with several thousand loyalists, out of the city.

4

The Trials of the
Republic of Pennsylvania

B Y THE SUMMER OF 1778, Paine was free to return to
Philadelphia. What he found in the liberated American
capital was a city struggling with war-time economic disruptions. Much of the energy and financing behind its prewar
industrial boom had dissipated as the city's commercial elite
turned their attention to various forms of war-time profiteering. Meanwhile, the city's artisans and small merchants
struggled to survive in the face of scarce credit and continued
interruptions to overseas trade. Paine's own circumstances
were similarly grim. With no residual income from *Common
Sense* or the *Crisis* essays, he was broke. A plan to publish a
compendium of his Revolutionary War writings fell victim to
a war-time paper shortage. For the first time since coming to
America, Paine would have to make his living with something
other than his pen.[1]

In November 1779, the Pennsylvania General Assembly, at
the time controlled by its most radical, most democratic fac-

tion, hired Paine as its clerk. In this capacity, he had a hand in drafting a series of remarkable measures, among them Pennsylvania's Act for the Gradual Abolition of Slavery, the first of its kind in the United States. But Paine was not content to bide his time as an obscure functionary. Nor was he content to abandon his revolutionary ideals. Committed as he was to the American cause, and firm as he was in his attachment to his new country, Paine never abandoned the Old World. It remained at once a source of deep anger and great hope. On the one hand, Paine despised its kings and its reactionary enemies of America's republican ideals. On the other, he saw in Great Britain promising signs of imminent revolution.

After a string of American military failures, including the fall of Charleston in May 1780 and the routing of the Continental Army in South Carolina that August, Paine began contemplating a scheme to return to England. Although he had become famous in his adopted country, he remained little known in the place of his birth. This relative anonymity, Paine convinced himself, would allow him to do what he had done at the beginning of his career in America. In effect, he could write a *Common Sense* for Britain. "I do not suppose that the acknowledgment of independence is at this time a more unpopular doctrine in England than the declaration of it was in America immediately before the publication of the pamphlet *Common Sense*," he told Nathanael Greene. Much as he had for the Americans, so he could persuade his former countrymen of the hitherto unthinkable: that the time had come for a radical reform of government.[2]

Greene warned that the risks were great. Discovery could mean a treason trial, a circumstance that grew more likely after the October execution of British Major John André for espionage. André had been secretly negotiating with the traitorous Benedict Arnold for the surrender of American soldiers at West Point when he was captured and brought to the American camp at Tappan, New York. In a gesture that infuriated British military officials, Washington refused to grant the British officer's wish to be executed by firing squad, and he was instead hanged as traitor and spy. With the British clamoring to avenge the execution of so honorable a military officer, an American traitor could now expect to suffer the fate of Major André.

In the end, Paine abandoned the project in favor of a less hazardous route abroad. In February 1781, he quit his job with the Pennsylvania Assembly to join Congressional emissary John Laurens and his secretary, William Jackson, on a financial mission to France. French military support had clearly changed the complexion of the war, but it had not brought the swift end patriots expected. If anything, the war was beginning to seem even more intractable, as the British government showed no signs of abandoning the fight and as the Americans struggled to fund their military. What Paine hoped to do for the delegation is not clear—but as an unofficial member, he paid his own way, a substantial expenditure for a man who had been living on a modest civil servant's salary. Perhaps he had planned to stay in France and propagate his revolutionary views from there. But for reasons that are also unclear, he returned with Laurens after only seven months away.

Having spent what money he had on the trip to France, the forty-four-year-old Paine was reduced to begging friends and acquaintances for food and money. It was all terribly humiliating, especially for a man so convinced of his own selfless dedication to the great cause of America. His good deeds, now including whatever role he played in Laurens's commission, which had yielded 2.5 million *livres*, were being repaid with indifference and contempt. Paine's patriot euphoria gave way to despair and he took little pleasure in the final American military victory at Yorktown. Less than two months after Washington's army achieved the surrender of General Cornwallis, Paine wrote the great American commander. He barely mentioned Washington's crowning achievement. Instead, he confessed that "it is seven years, *this day*, since I arrived in America, and though I consider them the most honorary time of my life, they have nevertheless been the most inconvenient and even distressing." His selflessness, Paine wrote, appeared to have been for naught:

> From an anxiety to support, as far as laid in my power, the reputation of the Cause of America . . . I declined the customary profits which authors are entitled to, and I have always continued to do so; yet I never thought . . . , but that as I dealt generously and honorably by America, she would deal the same by me. . . . I have only experienced the contrary.

What, Paine asked, was left but for him to leave the adopted homeland that had abandoned him? "In this situation I cannot go on," he explained to Washington, "and as I have no inclination to differ with the country or to tell the story of her neglect, it is my design to get to Europe, either to France or Holland.

I have literary fame, and I am sure I cannot experience worse fortune than I have here . . . after all, there is something peculiarly hard that the country which ought to have been to me a home has scarcely afforded me an asylum."[3]

Had he the means, Paine surely would have left America; but instead he made do, thanks mostly to Robert Morris, the Philadelphia financier and Superintendent of Finance for the Continental Congress. Morris was among the most controversial figures of the Revolution. Orphaned at a young age after his father's accidental death, he became a protégé of Thomas Willing, a mayor of Philadelphia and one of the city's leading merchants. As Willing pursued his public career, Morris took an increasingly prominent role in the business, ultimately becoming Willing's partner. The partnership flourished, coming into possession of ten ships and mercantile contacts around the Atlantic. Its diversified portfolio of business included real estate, finance, commodity trading, and projects to build slave plantations in Louisiana.

As independence approached, Morris became more active in public life and through the war served as a member of the Continental Congress and the Pennsylvania legislature. His proximity to power, immense financial resources, and extensive mercantile network also left him well placed to assist with the provisioning of the army, which he did to his great personal profit. After parting ways with Willing, that profit only grew and Morris turned his attention to one new investment after another. While his countrymen endured the hardships of war, Morris's businesses flourished. Some, including Paine, attacked him for profiteering. But few other Americans had the financial resources or the connections to acquire the arms and supplies the Continental Army so desperately needed. When Congress had to elect a chair of its Finance Committee, it thus turned to Morris. In the aftermath of a disastrous 1780, which included British victories in the south and another desperate, freezing winter for Washington's troops, Congress replaced that committee with a new quasi-executive Superintendent of Finance. The natural choice for the office, once again, was Robert Morris.

Morris's charge was among the war's most challenging. He had to restore Congress's credit and to find funds to pay the Continental Army, but he had to do so without one of the cornerstones of fiscal power. The Continental Congress had no direct powers of taxation. For most of its revenue, it relied on the states. As part of the agreement establishing the Articles of Confederation, the charter that defined the powers of Congress, Congress could requisition funds directly from the states. This quota system provided nowhere near enough to cover the costs of the war, leaving Congress deeply indebted to foreign governments and their bankers, as well as to the likes of Morris himself. As Congress's debt grew, the cost of borrowing grew as well. An ever more indebted United States, struggling to pay down its existing debts, allowed creditors to justify higher interest rates.

Morris hoped to stop this downward fiscal spiral by reducing the size of Congress's debt. But doing so meant short-term pain. The states would have to consent to a more regularized and centralized system of taxation. Morris's plan involved a new nationwide 5 percent duty on imported goods. In a country where taxation had already proven explosive, such a tax seemed the most politically feasible because it fell most heavily on those who consumed most. But any national tax, no matter how benign, would generate a political firestorm. If Morris were to have any hope of persuading his recalcitrant fellow Americans to accept this expansion of Congress's power to tax, he would need to take his case directly to the people. The financier Morris had little experience with the eighteenth-century media and its powers of persuasion. For that, he needed a spin doctor and there was none better than Paine. Morris hired Paine for a modest eight hundred dollars a year to sell his new tax scheme.

While Paine was clearly writing for hire, his views on public finance aligned with Morris's. Both men saw in the burgeoning government debt a path to ruin; both favored money redeemable in gold and silver specie, over the paper notes Congress and state governments had issued during the war; and both men saw taxation as a crucial instrument with which to reduce debt.

The clearest evidence for the consistency of the two men's views came in a supplementary *Crisis* essay that Paine published in October 1780, more than a year before he began working for Morris. The entire thrust of "The Crisis Extraordinary" centered on the problem of taxes. With Parliament's taxation schemes still on their minds, Americans took a principled stand against government revenue demands, but meanwhile the army fighting for their liberty suffered. It was a paradox Washington struggled against throughout the war. As he tirelessly reminded Congress, if the American army had any hope for victory, it would have to be clothed, housed, fed, and paid. Paine too had been disgusted by the troops' condition and took it upon himself to give his countrymen a lesson in the politics of taxation. They could accede to Congress's request, he explained, or they could suffer military defeat. The costs of the latter would far exceed those of the former. For what Congress was asking Americans to pay was paltry in comparison to what a victorious King George would demand.[4]

In the spring of 1782, Paine began writing a series of newspaper essays that made very similar arguments on behalf of Morris and the national treasury. Indeed, he took his case one step further, arguing that not only would American taxes be less onerous and more fair than their British counterparts, they were also a matter of national honor and duty. "It is a pity," he wrote in the *Pennsylvania Packet*, that there was not "some other word beside taxation . . . for so noble and extraordinary an occasion, as the protection of liberty and the establishment of an independent world." For what, after all, was he asking of his fellow Americans but that they share the costs of liberty?

Paine knew he faced an uphill fight. As he observed many times, he and his countrymen had just endured nearly seven years of war to free themselves from a state that had used tax revenue to finance war against its own subjects. The American national government, he assured readers, would be different. It would be transparent—government finances would be a matter of public record. Its revenue demands would be minimal—the absence of endless self-serving dynastic wars meant that most of the taxes could be used for the short-term goal of reducing the national debt, and in turn, serving a long-term objective of lowering taxes. But above all, "Government and the people do not in America constitute distinct bodies." Those who govern will thus be "subject to the same taxation [as] those they represent, and there is nothing they can do that will not equally affect themselves as well as others."[5]

Since Congress had no power to impose a national tax, Morris's new taxation scheme would have required the unanimous cooperation of the states. Without this, a single nontaxing state would undermine the whole system by drawing all imports to its ports. Thanks partly to Paine's pen, by the fall of 1782, it appeared that such unanimity was within reach. The one remaining obstacle was Rhode Island.

Hoping that the small state could somehow be brought around, Morris enlisted Paine to urge its citizens to change course. Paine made his case in a series of letters in the *Providence Gazette*. The gist of his argument was that the existing system of taxation was unfair. Almost every state in the new American Confederation relied for revenue on taxed property. This meant that farmers and small producers paid the bulk of the nation's taxes, and, with no tax on trade, merchants and consumers paid very little. Morris's new tax was intended to spread the burden more equitably. Rhode Island's merchants understood that they would bear the brunt of the new tax, and they knew that Morris's plan worked only as a national plan. Although the states would be responsible for enforcing the new imposts, because the plan demanded that they all do so, and because its design came from an agent of Congress, the plan effectively empowered Congress to tax, something far outside of its constitutional authority. If opponents needed a surefire way to scuttle the plan, this was it. They could simply point out that it put the United States on path to becoming a centralized British-style imperial state.

This was exactly what Rhode Island's influential merchants did. If Congress required more funds, they argued, it should raise the quotas and leave it to the states to determine how those quotas would be filled. The problem, as Paine, Morris, and most members of Congress knew all too well, was that the burden of added quotas would fall once again on farmers and other property holders, many of whom were already deep in debt. To tax them further would likely worsen the precise problem the import duty was meant to address. Farmers would respond to heavier taxes with demands for cheaper credit. State politicians would then print more worthless paper money, using inflation to lower credit costs, but leaving states unable to fulfill Congressional quotas.

In the end, Paine's arguments had little impact. Rhode Island's legislature refused to budge and Morris's scheme was abandoned. The fight left Paine acutely aware of the political fragility of his new country. Unless Americans could find some common ground on fiscal issues, their union was unlikely to survive, even without the burdensome costs of war.

Morris, dejected by the failure of his plan, resigned from his Congressional post, leaving Paine once again out of work. But in a rare moment of financial good fortune, Paine's pleas for some kind of government compensation were finally answered. As a reward for his service during the war, in the spring of 1784, New York granted him three hundred acres of confiscated loyalist farmland in New Rochelle, north of New York City. In addition, after Washington's persistent lobbying, in 1785 Congress awarded Paine an additional three thousand dollars. For the first time in his American life, he could enjoy a modest degree of financial independence.

WITH THAT INDEPENDENCE, Paine devoted himself to the internal improvement of his adopted country. In the aftermath of the war, the unifying powers of war-time crisis receded amid political acrimony, much of it driven by the kinds of fiscal disputes Paine had faced in Rhode Island. Paine came to see in this growing political division the greatest threat to the new nation's long-term survival. He was particularly disgusted with what was happening in Pennsylvania, a state that had already infuriated him with its earlier failure to provide for his financial welfare.

"Instead of that tranquility which [Pennsylvania] required and might have enjoyed, and instead of that internal prosperity which her independent situation put her in the power to possess," he wrote Daniel Clymer, state assemblyman from Berks County, "she has suffered herself to be rent into factions." The origins of this factious spirit were clear. The only real mystery was that farmers living in the western parts of the state, long accustomed to underrepresentation in the state assembly, maintained any allegiance to Pennsylvania at all. "They are not affected by matters which operate within the old settled parts of the state," Paine explained. For "they are not only beyond the reach and circle of that commercial intercourse which takes place between all the counties on [the east side of] the Susquehanna and Philadelphia, but they are entirely within the circle of commerce belonging to another state, that of Baltimore."

When whole regions of Pennsylvania owed their economic well-being to the city of Baltimore, could it be any wonder that their representatives had no interest in the collective economic good of the state? Indeed, "some of them may probably think that it would be no disadvantage to their situation if the Delaware through which all the produce of the Counties east of the Susquehanna must be exported, were shut up."[6] One newspaper correspondent noted in 1787, "Happy would it be for Pennsylvania if her boundaries were comprised by the Susquehanna; we should be more compact and more united."[7]

For Paine, the state's sectional divisions revealed a fundamental political problem. Self-governing political communities had to be bound together by trade. "In all my publications," he would later recall, "I have been an advocate for commerce, because I am a friend of its effects." Commerce sustained the social bonds, the common language, and the sense of mutual obligation that made republics possible. "It is a pacific system," Paine explained, "operating to unite mankind by rendering nations, as well as individuals, useful to each other." In 1795, promoters hoping to revive the Delaware–Chesapeake canal project made the same point. Commerce brought people together, whether between nations or within them:

Commerce between the inhabitants of different countries . . . is the surest means of uniting all mankind, in one happy bond of civilization, peace and prosperity. By commerce, in this enlarged sense of the word, "the whole world becomes, as it were, one single family." What nature has denied to the inhabitants of one climate, is supplied by what she has liberally bestowed on another; and the super-abundance of each becomes *common stock*. What commerce, considered in this view, is to mankind in general, by means of *foreign trade* and *external navigation*; she is, in a smaller degree, to particular states and societies, by means of *inland navigation* and good *roads*; whereby the produce of one part of the country . . . is easily exchanged for that of another.

With part of Pennsylvania's trade tied to the capital city and part to Baltimore, no such easy exchange could occur and Pennsylvania's commerce was denied its pacific effect.[8]

Pennsylvania's sectional division became even more frustrating for Paine as he faced his state's most acrimonious political controversy since independence. That controversy centered on the Bank of North America, a private bank chartered in 1781, with dubious constitutional authority, by Congress as well as the states of Pennsylvania, New York, and Massachusetts. The bank was controlled by associates of Robert Morris, one of whom, Thomas Willing, was its president. To the bank's defenders—and Paine was to be among the most vocal—it was a vital instrument for stabilizing the nation's currency and stimulating its economy.

As the sole repository for Congress and the three sponsoring states, the bank would be, by default, the government's bank. For the bank's shareholders, there could hardly be a more reliable investment. The bank's charter stipulated that Congress and the other chartering governments would treat it as lender of first resort. Interest from these government loans would provide dividends for the bank's shareholders. Congress would use credit from the bank to issue new notes convertible to specie held by the bank. To Paine and the bank's other defenders, this constituted a shrewd public-private partnership. Private investors would profit but the government would gain a hard-money alternative to the virtually worthless paper promissory notes—the "Continentals"—it had issued during the war. A more stable currency supply, not constantly sapped by inflation, would also lower Congress's long-term borrowing costs.

To a growing chorus of detractors in the Pennsylvania General Assembly, the bank's immediate fiscal benefits were immaterial. What was truly important was the political meaning of the bank. The unelected financiers who controlled the new bank would now dictate how and according to what terms the government could borrow money; they would control government-issued currency; and they would have charge of public revenues deposited in the bank.

Hailing mostly from the state's western counties, the bank's enemies wondered how such an arrangement could do anything but give control of government to Morris, Willing, and their small circle of fellow Philadelphia financiers. And how could those bankers act in the interest of anyone but themselves and the shareholders? "The government of Pennsylvania being a democracy," declared William Findley, a state legislator from Westmoreland County in Pennsylvania's far southwestern corner, "the bank is inconsistent with the bill of rights thereof, which says that government is not instituted for the emolument of any man, family, or set of men." And yet that seemed precisely what the bank was designed to do. Another opponent writing in the *Freeman's Journal* under the pseudonym Philadelphiensis wondered how hostility to the bank could be at all surprising "when we consider that the vast funds of this bank consolidate the weight and influence of its proprietors who are a regularly organized body, subject to no control of government."[9]

Paine had no patience for such arguments. Far from serving a minute moneyed interest, the Bank of North America functioned as any effective public work should: it mobilized private wealth for public good, and it did so according to the regulating terms of law. By generating loans and by stabilizing the currency supply, it was actually the most democratic of institutions, benefitting equally the farmer and the merchant. In fact, it was the opponents of the bank who were undemocratic. For in their opposition to so manifestly beneficial an instrument of public good, Paine wrote in one of a series of newspaper columns on the topic, they betrayed a dangerous devotion to their own interests:

> The opposition [to the bank] . . . has the appearance of envy at the prosperity of all the old settled parts of the state. The commerce and traffic of the Back Country members and the parts they represent goes to Baltimore. From thence are their imports purchased and there do their exports go. They come [to Philadelphia] to legislate and go there to trade. In questions of commerce, and by commerce I mean the exports as well as the imports of a country, they are neither naturally nor politically interested with us.[10]

In other words, Paine was arguing, opposition to the bank had less to do with political principle than commercial self-interest.

With parts of the state directing business to Baltimore, there could be little hope for a representative government acting on behalf of the common good.

Had the state's capital been in the west; had westerners enjoyed better and more proportional representation in the General Assembly; and had the state's financial power been less concentrated, perhaps the fight over the bank would never have happened. But Paine and his fellow bank supporters never admitted these truths. Instead, they clung to the optimistic belief that commerce, made possible by internal improvements, would reunite the state.

5

The Schuylkill and Its Crossings

O N THE MORNING OF APRIL 20, 1789, George Washington and his retinue left Frederick, Maryland, for Philadelphia. Washington traveled north along the western bank of the Schuylkill River atop a white horse, followed by a swelling procession of well-wishers. The retired American general was en route from his home in Virginia to New York City, where he would assume the presidency of the new United States. For the previous four days, Washington had been feted from Alexandria to Baltimore and from Wilmington to Chester. He would now cross the Schuylkill River and enter the nation's former capital, greeted by crowds of cheering Americans; the jubilation would continue into the night as revelers rejoiced beneath a sky bright with fireworks.

The day's most remarkable moment came at about noon when Washington crossed the floating bridge at Gray's Ferry, south of Philadelphia. The bridge had become a fixture of municipal life, functioning as a gateway to America's largest city and the chief overland connection to its southwestern hinterland. It was only fitting that the prosaic structure be made a worthy passage through which "His Excellency" (as the new president was called) would enter a city he once defended with blood and honor.

Charles Willson Peale, the great impresario, former Continental Army captain, and painter, had accordingly transformed the bridge into a marvel of republican kitsch. At each end, he erected a twenty-foot triumphal arch, festooned in laurel branches. The bridge itself he decorated with shrubs, and along one railing hung the eleven flags of the states that had so far ratified the nation's new constitution. Draped on the other railing was a flag of the state of Pennsylvania, the very flag that had recently returned with Captain Thomas Bell from America's first commercial voyage to India. At each end of the bridge, banners flapped in the wind, broadcasting the soaring proclamations: "Behold the Rising Empire" and "May Commerce Flourish."[1]

During Washington's inaugural festivities, Thomas Paine was in England preparing to build the new prototype of his own bridge. But he knew the Gray's Ferry bridge well. The same sorts of floating bridges had been built during the war at Middle Ferry and at Upper Ferry, north of the city. This one was probably made from remnants of one of the war-time Middle Ferry bridges. Now, eight years after the fighting had stopped, these artifacts of war remained, little-noticed but essential conduits of trade and transport.

THE MID-ATLANTIC REGION had bridges dating far back into the colonial era. In 1697, William Penn had ordered the construction of America's first stone-arch bridge, across Pennypack

Creek, north of Philadelphia. A somewhat modified version of the seventy-three-foot bridge, now known as the Frankford Avenue Bridge, remains in use to this day. Permanent stone and timber bridges had also been built across Ridley Creek and Big Elk Creek in Maryland, and across small waterways in eastern Pennsylvania and western New Jersey. These bridges tended to span no more than a hundred feet, typically resting on midriver piers. For waterways of the breadth of the Schuylkill, ferries had always been preferable. Few colonists had the skills or the funds to bridge rivers three hundred to four hundred feet wide.[2]

There were legal barriers to surmount as well. The most formidable were posed by the charters and lease agreements colonies and municipalities used to establish complex public works. Ferries, for instance, were typically established through leases that stood for generations. In addition to granting ferry operators the right to charge tolls for their services, the leases gave them the privilege of doing so without any immediate competition. Since many colonial roads led to ferry piers, building bridges beyond the competitive reach of the ferries would have required the added cost of new roads, which was prohibitive. The result was that colonists rarely challenged ferry leases.

The arrangement was fundamental to early-American public works. Much like the charters and patents establishing some of the American colonies themselves (not to mention canals, turnpikes, bridges, and assorted other enterprises in England), these legal arrangements were designed to wed private gain to public good. By affording the ferrymen exclusive rights and revenues for the transit of goods and people across rivers, government addressed a public need at very little direct cost.

The Revolutionary War subjected these arrangements to a form of creative destruction. Faced with the prospects of British invasion, Pennsylvania's capital could ill afford the limitations of its old ferries and began replacing them with more efficient floating bridges. Rather than confront the tangled legal matter of who actually owned the new bridges, Philadelphia city officials allowed ferry leaseholders to take charge of the bridges as war-time necessity receded. In effect, ferrymen came to own Philadelphia's bridges and the toll revenue they generated.

In addition to their practical utility, these new bridges were safer than the other options. In early America, rivers were places of horror. In 1784, the *Pennsylvania Packet* reported a particularly disastrous accident on the Ashley River outside of Charleston:

> A Mr. Frazier, with 72 negroes belonging to Mr. Thomas Elliot, and a negro and horse, the owner not known, and a negro boy belonging to Mr. Frazier, were crossing the river, nearly in the middle, the boat separated in two, by which 48 of Mr. Elliot's negroes, the negro and the horse, together with Mr. Frazier's boy, were drowned, and Mr. Frazier (the owner of the ferry) very narrowly escaped.

In the north, winter ice heightened the hazards. John Hall, an immigrant mechanic who would help Paine build his bridge, observed that what Washington's army did in December 1776, Philadelphians did routinely, often paying with their lives. "They will pass with Boats," Hall wrote shortly after arriving in America, "when there is A Great Quantity of Ice in the River Steering the Chasm between when it is either going up or down with the

Tide and I am told they will Haul out their Boat and draw it over Sheets of Ice and then put it into the Water again." For some the voyage was fatal: "A man lost his Wife So . . . meeting a Woman to whom he was bewailing his loss . . . Shee God Bless hir hapend to have lost hir Husband in the Same manner." On a 1788 journey from New York to Philadelphia, the Frenchman Brissot de Warville summarized the problem well: "There is no doubt that sooner or later bridges will, wherever possible, replace these ferries, which are often dangerous. I was close to death once on one of these ferries, while crossing the Hackensack River."[3]

FOR ALL THEIR obvious benefits, however, Philadelphia's new bridges created one serious problem. Their operation and maintenance costs far exceeded those of other public works— at least in the short run. Roads, dams, and fences required little maintenance and for this reason were generally left under public control. They could be maintained by towns and localities through old systems of compulsory labor and road taxes. Depending on the scope of their property, able-bodied free men would contribute labor or funds to the maintenance of these works.

But bridges required carpenters and perhaps stonemasons— skilled tradesmen who could not always be found. Even when available, bridge builders and bridge operators faced challenges little known to proprietors of other public works. Canals and turnpikes rarely required complete reconstruction. But for bridges, floods and careening ice made investment on this scale routine. Floating bridges were particularly vulnerable, as British military engineers had discovered shortly after completing their own bridge across the Schuylkill. In October 1777, Captain John Montresor of the British Royal Engineers confided in his journal that "at 2pm the floating Bridge at Middle Ferry was carried down the Schuylkill by the N.E. Stormy High tide and rapid stream." The bridge had been in place for less than a month.[4]

More permanent bridges, built across midriver piers, fared little better. Sullivan's Bridge was built by General John Sullivan across the Schuylkill near Valley Forge in the winter of 1777–1778, on orders of George Washington. The bridge was a replacement for the floating bridge that had carried Washington's soldiers across the river to their famous winter encampment. The new structure, Washington hoped, would withstand the trials of late-winter ice and spring floods. For its builders, the bridge would also stand as a monument to their leader and the triumphs of patriotism. Captain Thomas Anburey, a British prisoner of war who crossed the bridge in 1778, imagined that "it was the intention of the Americans that this bridge should remain as a triumphal memento, for in the centre of every arch is engraved in the wood the names of the principal generals in their country; and in the middle arch was General Washington's with the date of the year this bridge was built."

As it turned out, within months of its construction, the bridge began showing signs of failure. Wood rot and winter ice had seriously damaged the road deck and supporting piers. Pennsylvanians had nonetheless grown accustomed to the bridge. In the fall of 1778, the state assembly commissioned John Edwards, a Philadelphia Militia officer, to estimate costs for repair; however, by the time Edwards received his orders,

Sullivan's Bridge had been entirely destroyed by winter ice. "Was it Repaired," Edwards told the assembly, "it [would] stand but a short time" before succumbing again to the forces of nature.[5]

Even when bridges were not completely swept away, they required constant maintenance. Rain, snow, ice, metal-clad wagon wheels, and horse hooves eventually turned the most robust wooden bridges into feeble, rotting hazards. General Washington himself discovered the peril in September 1787 as he returned to Mount Vernon from the Constitutional Convention in Philadelphia. After a rainy night in Wilmington, Delaware, Washington set out for Maryland, intending to cross Big Elk Creek at the town of Head of Elk, in northeastern Maryland. On approaching the creek, Washington determined that rains would make it impossible to traverse the nearby ford. "Being anxious to get on," he chose to cross "on an old, rotten & long disused bridge." The results were a near national calamity. While crossing the bridge with his two horses and carriage, part of the bridge gave way. One of the horses fell fifteen feet from the bridge, while the other narrowly avoided doing so and pulling with it Washington's carriage and baggage. Fortunately, the general himself was unharmed, but the story captivated the nation. It was reported in at least forty-six newspapers, from Georgia to Vermont.

If for most the accident was emblematic of Washington's godlike capacity to overcome life's perils, for some it was a reminder of the new nation's most treacherous hazard. Even with bridges, crossing rivers was life threatening. A decade after Washington's accident, the *Gazette of the United States* reported "numerous are the accidents that occur" on Philadelphia's floating bridges. "The week before last a carriage and horses were precipitated into the water from the bridge over the lower ferry," and the horses lost, the passengers nearly drowned. The accident was entirely owing to "the railing on these bridges, [which] either from the form of their construction, or from carelessness, affords very feeble protection."

War-time inflation added to the problem of maintaining Philadelphia's bridges. Ferrymen operated the former military bridges by the terms of their old ferry leases. These had fixed the tolls back in the days of the ferries. Now, with their toll income ravaged by inflation, bridge operators could rarely cover their costs, let alone maintain the bridges. Whatever benefits the old leasing system had provided for ferry operators had been eroded by the realities of the bridge business. Before long, the city of Philadelphia and the state of Pennsylvania found themselves faced with a failing system of public works.

The situation became so dire that in the spring of 1779, the state assembly raised toll rates. Nevertheless, inflation continued to erode leaseholders' incomes, and margins were too slim to account for any substantial damage to the bridges. As one Pennsylvania official observed, "the toll for Crossing the bridge over the Schuylkill, from the depreciation of the money, is become so trifling, that the [leaseholder] there assures me it is with difficulty he can support his family." He went on to warn that "should any accident happen to the Bridge it would be impossible . . . to maintain boats for any length of time at the present low rates of the ferriage." In other words, given the depreciation of American money, current toll rates would not even be enough to sustain ferry service should a bridge fail.[6]

The kinds of problems Philadelphia's bridge operators faced

were not universal in the new United States. In 1786, on the eleventh anniversary of the Battle of Bunker Hill, a group of investors introduced the Boston public to the first bridge across the Charles River. By early American standards, the bridge was a massive public work, some fifteen hundred feet in length and costing more than fifty thousand dollars. The bridge was built atop seventy-five oak piers, and at its center was a drawbridge "opened by a most ingenious machine which is moved so easily that two ten-year old boys can operate it." The new Charles River Bridge would remain in place for decades. By 1792, it had given rise to an imitator directly linking Boston to Cambridge.[7]

The happy story of the Charles River Bridge was largely a function of the Charles River itself. Unlike the Schuylkill, Susquehanna, or Delaware, the Charles passed through a tidal plain that constituted most of what is now Boston's Back Bay. By the time the river reached Charlestown, it had disgorged itself into these plains, leaving behind floodwaters and spring ice. Relative to even the Schuylkill, the Charles also has a modest watershed, a little more than 300 square miles. With its short length of about 80 meandering miles (the direct distance from the river's mouth to its source is only about 26 miles) and a fall of only 350 feet, floodwaters and ice floes on the river were unlikely to reach the velocity they did on the Schuylkill, a 130-mile long river, with a watershed of nearly 2000 square miles and a fall of nearly 800 feet. With no substantial marshes or swamps to absorb winter and spring freshets above Philadelphia, the Schuylkill routinely overflowed its banks, carrying away much in its path.[8]

In addition to its natural advantages, the Charles River

Bridge enjoyed a substantial financial advantage. Although the bridge was much longer and more costly than the floating Schuylkill bridges, its operators could expect something not possible on the Schuylkill. During the colonial era, as Philadelphia's hinterland expanded to the north, west, and south, the colonial government allowed for the creation of three ferries, leaving travelers a range of options for crossing the river. For travelers crossing from Boston to Charlestown, however, there was only a single option. As a means of financing the new Harvard College, in 1640 the Massachusetts Bay Colony granted the college exclusive rights to ferry goods and people between Boston and Charlestown. That concession became college property. Virtually all traffic between Boston, Charlestown, Cambridge, and other areas to the north used the Charles River ferry. The college jealously guarded the asset, challenging an earlier bridge proposal as an intrusion on its rights to the profits from ferry traffic.

After the Revolutionary War, with no British authorities to enforce ancient charters, the efficiencies of a bridge became hard to dispute and the state granted a group of private investors a forty-year lease for the new bridge. Although the leaseholders would have to bear the costs of building and maintaining the bridge and would have to compensate Harvard College for its lost revenue, they would otherwise enjoy a complete monopoly on traffic across the Charles—at least until 1792, when the state authorized the construction of the second bridge, to Cambridge. Even with the competition, the builders of the Charles River Bridge realized handsome gains. By 1805, an investment in the original bridge had risen in value by more than 300 percent.[9]

Philadelphians would enjoy no such gains from the bridge business. The natural and financial circumstances were simply too challenging. But nor could they afford to ignore their failing infrastructure. The city continued to struggle with the loss of trade to Baltimore. Without some measure to improve access to Philadelphia's markets, Baltimore would continue to prosper at its expense.

6

The Schuylkill Permanent
Bridge Company

I NSPIRED BY THEIR COUNTRYMEN in Boston, a group of prominent Philadelphians undertook to solve the city's bridge problems. In place of deteriorating and costly floating bridges, they proposed a new permanent bridge. And in place of outdated ferry leases, they sought a charter from the General Assembly, entitling them to sell shares in a new bridge company. Capital from the sale of shares would finance the construction of Philadelphia's first permanent bridge.

The scheme was brought before the assembly in November 1786 by the Philadelphia Society for Promoting Agriculture, a new learned society founded the year before. With a membership that included Robert Morris, Benjamin Rush, Benjamin Franklin, the lawyer and jurist James Wilson, the visiting English merchant and gardener Samuel Vaughan, and George Clymer, cousin of Paine's friend Daniel and heir to a mercantile fortune and member of Congress and the Pennsylvania General Assembly, the society asserted its considerable political and financial clout on behalf of the proposed bridge company.

By the fall of 1787, it had accumulated enough prospective investors to begin planning for the new bridge. The remaining barrier was the state government's consent. The Agricultural Society expected the bridge to be financed by Philadelphia's wealthiest, many of whom were its members (and also members of the state assembly), but there would be no way to pool citizens' capital without the government's legal sanction. By late 1787, this seemed imminent. The people's representatives had agreed to submit a proposed Act of Incorporation for the Schuylkill Permanent Bridge Company to public review.[1]

The scheme was a classic public–private arrangement. After the company chose a design and built its bridge, it would charge tolls sufficient to provide investors "with legal interest, and a reasonable gratification for the risque incurred on account of said Bridge." Once the company's debts were paid and its shareholders compensated, the state assembly would have the power to make the bridge free to the public. To the Agricultural Society's members, it was the most elegant solution to a pressing problem. But politics intruded. For western Pennsylvanians, the Permanent Bridge Company seemed like another Bank of North America—a boondoggle designed to enrich Philadelphia at the expense of the rest of the state. Opponents also questioned the General Assembly's jurisdiction in the matter. There was nothing in the state constitution that gave the assembly power to negate ferry leases granted by the old colonial government or the defunct municipal corporation that had long governed Philadelphia. In chartering a new bridge company, that was exactly what the legislature would be doing.[2]

IN 1786, THE YEAR the Agricultural Society began campaigning for a permanent bridge, Paine was consumed with the fight over the Bank of North America. In February, he published "A Dissertation on Government; the Affairs of the Bank; and Paper Money," followed by a series of newspaper essays defending the bank and attacking a new paper-money bill recently passed by the state assembly. Although there is no evidence that Paine was writing for hire, anonymous critics immediately questioned his motives. Just a few months earlier, in the fall of 1785, after all, Congress had paid him three thousand dollars. The attacks infuriated Paine, and may explain why he began abandoning public life for a life in architecture.

When Paine described the early genesis of his bridge in his 1803 "The Construction of Iron Bridges," he said nothing about any controversy over the bank. His ambition was simply to devise a bridge that could endure the vagaries of the American climate: "By violence of floods and breaking up of the ice in the spring . . . bridges depending for their support from the bottom of the river are frequently carried away." In exposing them to careening ice and floodwaters, the midriver piers on which most permanent bridges rested were the prime vulnerability of American bridges. Paine's solution was a single arch "that might, without rendering the height inconvenient or the ascent difficult, extend at once from shore to shore, over rivers of three, four or five hundred feet and probably more." In addition to lifting the bridge above the hazards of rushing ice and water, Paine believed his design would improve river travel by leaving "the whole passage of the river clear."[3]

In order to build a practical single-arch bridge, Paine had to overturn centuries of architectural wisdom. Bridges, roofs, viaducts, arcades, and other spans were typically supported in one of two ways. They used the post-and-lintel method, with flat beams resting on columns. This was what the builders of the Charles River Bridge had done. The other method, more common in Europe, relied on semicircular masonry arches. Roman architects had discovered that the most effective means of distributing a span's load was to direct weight outward and downward in equal measure. A stone or brick arch made from half of a circle provided the perfect ratio of lateral and vertical distribution of weight.

For Paine, this bit of ancient architectural wisdom presented a difficult problem. The height of a semicircular arch was roughly half its width. A single arch across the Schuylkill would thus be some two hundred feet tall. In order to bring traffic across such an arch, a series of enormous and cumbersome structures would have to be built. The bridge would require spandrels, structures filling the angular void between the curve of the arch and the flat road deck. These were normally built from a combination of loose aggregate fill and masonry, all adding to the weight bearing down on the arch. To bear this added weight, the arch itself would have to be stronger. Similarly, in order to bring traffic to the elevated road deck on the top of the arch, builders would have to construct embankment ramps and towers. For a two-hundred-foot arch, these would be enormous and costly. To avoid these kinds of added expenses, Old World bridge architects, with very few exceptions, relied on a sequence of small masonry arches. The Ponte Sant'Angelo, built across the Tiber in Rome during the reign of Hadrian, for example, consists of five stone arches, the longest of which is about sixty feet.[4]

The Westminster Bridge in London—which upon its completion in 1750 became the first masonry bridge built across the Thames since the original London Bridge was finished more than five hundred years earlier—consisted of fifteen consecutive semicircular masonry arches. This bridge illustrated precisely the kinds of problems Paine hoped his single arch would avoid. Although river ice was seldom a problem in London, bridge piers were still vulnerable.

As a tidal river, the Thames rose and fell by as much as fifteen feet during the course of a day. Any large obstructions in the river could produce furious eddies and rapids as the river water rushed in and out to sea and could even cause the river to overflow its banks. A new bridge, with a series of new piers, promised to make the problem worse. Charles Labelye, a Swiss-born mathematics instructor, military engineer, and builder of the Westminster Bridge, was the first bridge architect to attempt to address the issue. In the late 1730s, as he was finalizing designs for the Westminster Bridge, he recognized that by measuring the fall, or the difference between the water levels above and below a bridge, he could calculate the relative disruption caused by piers of varying sizes and shapes. Using this method, he also discovered that the London Bridge, long the bane of Thames riverboat men, could produce a tidal fall of as much as six feet. The massive stone piers supporting the bridge occluded nearly three-fourths of the river's flow.

By minimizing the width of supporting piers, Labelye concluded, the fall could be reduced to as little as several inches.

But narrowing the piers could be accomplished only by minimizing the size and weight of the arches above. This would, in turn, limit the size of the boats that could travel upriver, beyond the bridge. The largest of the Westminster Bridge's fifteen arches, the central arch, was seventy-six feet in width and about thirty-eight feet high, large enough to accommodate Thames barges, but not high-masted sailing ships.

Even with Labelye's innovations, the new piers presented problems. Somehow these masonry structures would have to be built, submerged, and fixed in place on the river's bottom. The method Labelye settled on involved eighty-foot "caissons." These were wooden barges upon which masonry for the piers could be set before the whole was sunk to underwater foundation pits. Labelye had calculated that this method, in the service of relatively narrow semicircular stone arches, would provide sufficient stability for the bridge. Alas, he was wrong. In late 1747, before construction of the Westminster Bridge was finished, fast-moving water had eroded the river bottom and one of the piers began to settle, damaging two adjoining arches. Similar problems persisted in the century after the bridge was completed, until finally, what remained of the whole crumbling pile was demolished in 1862.[5]

Paine's solution to all of these problems came from a simple observation. For a single-arch bridge, a "small segment of a large circle was preferable to the great segment of a small circle." Instead of building his bridge from one-half of a circle, Paine's arch would be fashioned from a much smaller fraction of a much larger circle. With its lower arc, cumbersome and expensive spandrels and embankment ramps would be unnecessary. In effect, the load-bearing component of the bridge and the deck of the bridge could be merged into a single, elegant structure. The problem Paine faced was that this proposition flew in the face of centuries of architectural wisdom. The balance of downward and outward forces made possible by semicircular arches would have to be achieved in some new way. Paine came to believe that the answer would lie in the configuration of lighter, stronger materials.[6]

His conclusion came from nature, which increased "the strength of matter by dividing and combining it, and thereby causing it to act over a larger space than it would occupy in a solid state, as is seen in the quills of birds, bones of animals, reeds, canes, etc." Each of these natural members was hollow and light, yet exhibited strength that seemed far out of proportion with its mass. The principle, Paine would repeatedly emphasize, was best expressed in "the spider's circular web, of which [the shallow arch] resembles a section." Like the spider's web, Paine's shallow arch would have strength that seemed impossible for so light a structure.[7]

IF PAINE'S IDEA was to be anything more than a curiosity, he would have to do what architects of buildings, ships, and other structures had done for centuries: he would have to demonstrate its validity with a precisely scaled model. But building that model would be no simple matter for Paine. Beyond his days as stay-maker, he had never really made anything. What he needed was a master model builder. Although Philadelphia abounded in skilled craftsmen, few possessed the range of abilities and the tools needed to fashion so precise and sturdy a thing as a model bridge built perfectly to scale.

Paine was fortunate to meet the very man for the job in the person of a consumptive fifty-year-old English immigrant. John Hall had arrived in Philadelphia in August 1785, leaving behind a family farm and gristmill in Leicester, not far from the booming industrial town of Birmingham. Hall came to America intending to buy land and resume his former trade as small farmer and miller. But he needed temporary employment before establishing himself in the countryside. Through mutual friends, Hall and Paine met in November, and within weeks Hall had begun working for Paine on a thirteen-foot wooden model of a bridge. Before immigrating, Hall had led the life of a country miller, but he was remarkably learned. "Asthmatic" episodes, fevers, and bouts of nausea, all caused by his tuberculosis, fed an obsession with medical science. The mechanics of maintaining his family's gristmill fed his fascination with nature's mysteries. He studied the writings of Isaac Newton and countless other natural philosophers. He also read such well-known political philosophers as Thomas Hobbes and the French Baron de Montesquieu. He owned a collection of deist tracts and books by such British radicals as Richard Price, the Unitarian opponent of the American War.[8]

But what made Hall so valuable to Paine was less his taste for medicine, science, and philosophy, or his progressive religious interests, than his equally well-cultivated comprehension of the mechanical arts. Before coming to America, Hall had been a maker of things. When not tending his Midlands gristmill, he spent his days and nights fashioning what was essentially miniature furniture. From walnut, mahogany, and other hardwoods, he made delicate boxes with intricate inner compartments and drawers, small precise hinges, locks, and keyholes. His customers purchased these to hold mechanical tools, scientific instruments, books, jewelry, glassware, guns, personal papers, and other valuables. In the parlance of nearby Birmingham, Hall had been what was known as a "toy-maker." Such a person did not make toys for children, but instead made small trinkets, including buttons, shoe buckles, sword hilts, watch chains, candlesticks, and the kinds of delicate small objects Hall made. So well-known had the Birmingham area become for its toy-making that in a 1777 address to Parliament, Edmund Burke called it the "toy-shop of Europe."[9]

The city's best-known toy-maker was Matthew Boulton, who, in partnership with the Scot James Watt, became Great Britain's foremost manufacturer of steam engines. Boulton and Watt were also part of an extraordinary learned society that arose in Birmingham. The Lunar Society attracted the region's most inventive minds, including Erasmus Darwin, the physician (and grandfather of Charles); Joseph Priestley, the freethinking chemist and champion of American independence; the potter Josiah Wedgwood, father of British mass production; and of course Boulton and Watt, whose steam engines revolutionized the British iron and coal industries.

FOR MOST OF HIS ADULT LIFE, Hall had spent his time making boxes, reading books, and tending his mill. In America, he broadened his horizons. When not crippled by tubercular fits, he spent his days studying the workings of his new country. Hall sat in the gallery of the Pennsylvania Statehouse observing court proceedings and assembly debates. He discoursed with his new American friends about the nature of law, government, and the mechanical trades. He trolled the streets of Philadelphia studying the terms of trade and business, recording the peculiar habits of the Americans. And what he found, he quite liked.

The United States "is the best Industrious mans Country Known," he explained to his nephew Joseph, because here, one's skills and one's labor were truly one's own. "A Prudent Industrious Farmer may live well and Comfortably and have neither the fear of a frowning Land lord nor the fruits of his Industry Stolen from him by the proud & Imperious priest," he noted. But what really made America so remarkable was its people's unmistakably democratic ways: "You may as well Expect Grapes to Grow on thorns as to hear them Say master or Mistress[.] No those words wd Choke them." Everyone in America, or at least in Philadelphia, it seemed, was treated equally.[10]

Hall's relationship with Paine echoed these American ways. Although Paine was the architect and Hall the model builder, the two were fundamentally collaborators. Hall discussed design decisions with Paine, and Paine assisted Hall with the model building. The collaborators also shared the duties of interpreting their models to an interested public. Here too, the democratic principles Hall so admired would govern the process of design and invention. What made models so effective was the universality of the language they spoke. Any reasonable, sensitive viewer could judge a model's beauty and integrity. This was particularly true of Paine's thirteen-foot bridge, since anyone could test the design simply by walking across it.

7

The Magical Iron Arch

ONE JANUARY EVENING IN 1786, after a round of cards, Paine invited a group of visitors up to the Philadelphia garret housing Hall's workshop. "The ladies were handed over [the bridge], one by one." The structure, made from twelve-inch cherrywood spars, assembled into a series of parallel supporting ribs, rose from the floor a little less than six inches. As the ladies traversed its gentle span, the small wooden bridge appeared to support their weight with integrity and ease. Paine was thrilled by this demonstration and the following day invited Philadelphia's greatest mechanic, the scientific instrument maker David Rittenhouse, to view his model.

Although Rittenhouse was impressed, he did observe a potential problem. While bearing visitors' weight, the bridge sagged slightly, causing its feet to slide outward. This lateral thrust, Rittenhouse suggested, caused by the downward pressure of weight on the arch, would slowly weaken the bridge. Rittenhouse's fear was that structural deterioration, combined with the invariably damaging effects of the American climate, would lead to catastrophic failure. The solution, he advised, were abutments whose mass was sufficient to counter the compression at the arch's center. The point was important, though hardly damning. For Paine, however, it was a very hard knock.

Among the virtues of his design, he believed, was its elegance and apparent lightness of structure. The idea that ever-larger spans would need to be countered by ever-larger abutments, fashioned from timber or stone, would add to the cost of the bridge, perhaps outweighing the benefits of a single arch.

Paine and Hall had chosen to build their first model from wood, America's cheapest, most abundant building material. So prevalent was wood construction, in fact, that some worried about the sustainability of American building practices. Peter Kalm, a Swedish naturalist who had traveled in the American colonies from 1748 to 1751, feared for the fate of the white cedar tree, whose wood colonists used to shingle their roofs. In New Jersey, where he observed the harvest of the trees, colonists seemed "bent only upon their own present advantage, utterly regardless of posterity. By this means many cedar swamps are already quite destitute of cedars, having only young shoots left."[1]

Wood's structural strength lies in its flexibility or its capacity to absorb tension caused by twisting and pulling. It is limited, however, in its capacity to absorb a compressive force produced by dead weight. To compensate for this limitation, Rittenhouse's observation suggested that any wooden arch would require abutments sufficient to counter whatever downward force its load imposed. Stone, the other principal bridge-building material, has the opposite qualities. Its capacity to resist compression is far greater than its capacity to endure the twisting and pulling associated with tension. Built from stone, then, a long, low-slung arch of the sort Paine envisioned would be just as vulnerable to collapse as a wooden arch, but for the opposite reason. Any twisting or pulling forces could dislodge stones and cause failure. So it was that, in their vulnerability to the contrary forces of compression and tension, wood or stone bridges built from a small segment of a large circle remained an architectural rarity.[2]

In January 1781, British builders unveiled an ingenious alternative to stone and timber construction in the form of a stunning iron bridge erected over the Severn River Gorge in the English West Midlands. Earlier in the eighteenth century, Sir Christopher Wren had used iron columns to support the galleries in a refurbished House of Commons, and architects in the iron-producing regions of England had begun using the same technique to support church galleries. But never before had an entire structure been built from iron. The unique bridge was simply called the Iron Bridge, the name also of the village that arose at its northern gate.

Travelers from around the British Isles and Europe came to marvel at the extraordinary structure. For painters and engravers, the one-hundred-foot iron span would be profitable subject matter for decades to come. Its image would emblazon dishware, trade tokens, bank bills, business cards, and other industrial-age ephemera. While serving as American minister to France, Thomas Jefferson purchased a print depicting the bridge; the image would later hang in his dining room at Monticello next to another depicting Virginia's Natural Bridge.[3]

The appearance of this architectural marvel near the Shropshire town of Coalbrookdale was the result of the area's unique industrial history. Since the reign of Elizabeth I, brick makers, potters, lead smelters, and other small manufacturers had been coming to the Severn River Gorge to exploit its abundant coal reserves. In 1709, the Quaker Abraham Darby, a Coalbrookdale ironmaster, discovered a means of using that same resource for transforming raw iron ore into iron. For centuries, the smelting process, which extracted impurities from iron ore to make pig iron, used charcoal for fuel. But the high cost of the wood needed to make the charcoal meant that iron was too costly for use in anything so extensive as a building or a bridge. Darby discovered that after slowly cooking coal, the by-product (known as "coke") could replace charcoal in the iron-making process and could do so at lower cost while producing greater purifying heat. The discovery transformed Shropshire into England's leading iron producing region. By the end of the eighteenth century, the booming iron industry had left its hillsides scarred by coal and iron mines.

As the region's economy grew, the Severn, which had been the crucial link between Shropshire and the rest of England, had become something of a barrier. Local officials, small merchants and farmers, Quaker worshippers, and others whose affairs took them across the river, faced a grave challenge. The river's banks lay at the bottom of a deep gorge, parts of which rose nearly vertically. Making the journey across the river by foot or on horseback was difficult enough, but for carriages and carts hauling coal or iron ore, it was often impossible. By the 1760s, the difficulty of crossing the Severn had become a drag on the local economy.

The scheme to confront this problem with a new iron bridge came from Thomas Farnolls Prichard, a prominent architect from Shrewsbury, just upriver from Coalbrookdale. As his architectural reputation grew, Prichard had become interested in bridges and in 1773 built a large masonry bridge at the convergence of the Severn and the newly completed Staffordshire & Worcestershire Canal. The same year, Prichard began designing a bridge for the Severn Gorge near Coalbrookdale. With the encouragement of Shropshire iron-masters John Wilkinson and Abraham Darby III (grandson of the first Abraham Darby), he began exploring the use of iron for this new bridge. John Wilkinson had become iron architecture's strongest advocate in Britain, and was said to have requested that he be buried in an iron coffin. Darby was less well-known and would be less prosperous than Wilkinson, but his Coalbrookdale ironworks were perfectly situated to fashion the members of a new iron bridge. In October 1775, Darby was commissioned to cast the bridge and six years later, this industrial-age marvel was completed. Pritchard died in 1777, before construction began, so it was left to Darby to build the structure.

Although the cast components of the Iron Bridge weighed as much as six tons, the total weight of the bridge was much less than that of a comparable stone bridge. Its cost, although higher than expected, was also lower than a stone-arch bridge. Much of this was owing to the relative ease of construction. Darby used derricks anchored to the banks of the river to lift five partially assembled cast-iron ribs into place. Within just a few months he had erected the entire iron arch without any substantial interruption of river traffic. It was a miracle in construction, performed by a group of ironworkers with no formal architectural or engineering background.

NEITHER HALL nor Paine left any mention of the Iron Bridge, even though Hall, who hailed from a region not far from Shropshire, surely knew of the structure. But in a fashion typical of Philadelphia, Birmingham, and other provincial centers of eighteenth-century innovation, the two men arrived at a conclusion not all that different from innovators on the other side of the Atlantic. They determined that iron construction would redeem Paine's design. Iron was less vulnerable to tension than stone but was more rigid than wood. With the structure itself absorbing the lateral thrust, massive embankment towers would be unnecessary.

In the spring of 1786, Paine and Hall began building a new cast-iron model. Upon its completion in June, they displayed it at Franklin's house, an obvious place to attract Philadelphia's investors. After seeing the bridge, Gouverneur Morris, the New York attorney who now lived in Philadelphia and would become one of Paine's sworn enemies, encouraged Paine to consider building an iron bridge over the Harlem River near Morris's family's estate, Morrisania, in what is now the South Bronx. While nothing came of the Harlem River bridge, Morris and his half brother Lewis were sufficiently interested in Paine's bridge to convince him that he might find in New York City the same interest he found in Philadelphia. Paine and Hall followed Morris's counsel and, by carriage and cart, took their model to New York.

The trip allowed Paine to check on his New Rochelle property and Hall to visit his nephew, Jack Capnerhurst, an ironworker who had immigrated to Morristown, New Jersey. During the summer of 1786, Hall and Paine began to consult with Capnerhurst, who had worked for Boulton and Watt before coming to America. These conversations gave Paine and Hall a subtler understanding of the properties of iron.[4]

AFTER RETURNING from New York, they built a third, much stronger and more stable model. Capnerhurst had urged them to use new iron-casting methods perfected in England, but the architect and his model builder could find no American iron founders skilled in those methods. Instead, they used wrought iron, which had to be forged by hand to expel the impurities that could make iron brittle. The process meant that instead of fashioning the bridge from the kinds of large castings used in the Iron Bridge, they would build it from a lattice of small hand-wrought iron bars.

These bars, assembled into individual arched ribs and then tied together to achieve the majestic whole, Paine came to believe, constituted another of his contributions to bridge architecture. By building bridges from many small parts, rather than a few enormous ones, bridge construction became a more adaptable enterprise. Now, instead of the unique product of a provincial iron industry and its industrial produce, an iron bridge could be an exportable commodity and could be erected wherever the need for such a structure arose. "Among the advantages of this construction," Paine wrote, "is that of rendering the construction of bridges into a portable manufacture, as the bars and parts of which it is composed need not be longer or larger than is convenient to be stowed in a vessel, boat or wagon."[5]

For now, though, Paine's design would make for a much more laborious model-building process. Hundreds of small metal bars would be assembled and connected tinkertoy-like, using nearly four thousand temporary wooden blocks. The bars would then be riveted together to form the structural ribs. The process of raising the ribs and tying them together demanded temporary scaffolding to hold the ribs in place, while a wooden floor and wrought-iron banister were added. It was a cumbersome and costly undertaking. In addition to Hall, Paine had to employ an iron forge to fashion the metal bars and a joiner to build the wood flooring for the bridge.

Through November and much of December 1786, Hall worked furiously, all the while struggling with crippling night sweats, coughing fits, nausea, fever, and the effects of the opium with which he treated his tuberculosis.[6]

PAINE, MEANWHILE, had a new sense of urgency. While Hall worked to finish the model, the Pennsylvania assembly was debating a proposed charter for the Schuylkill Permanent Bridge Company. In an effort to persuade legislators of the feasibility of a new bridge, the Agricultural Society displayed another architect's bridge model at Carpenter's Hall, just a few blocks from the state capitol. The model, built by a tailor named Sellers, infuriated Paine, who thought it violated the fundamental premise of any worthy American bridge design: it used midriver piers. Fearing that the state assembly would cave to pressure from the Agricultural Society and endorse such a design, Paine wrote George Clymer, who had promised funds for the new bridge project, pleading for patience and urging him to exert whatever influence he could to delay any final decision. The bridge then being considered, Paine explained, was not only unlikely to stand, "it is also a matter of more hazard than [the Agricultural Society] are aware of." The supporting piers would create "obstructions [in] the bed and channel of a river," and since "the water must go somewhere—the force of the freshets and the ice is very great now but will be much greater" when constrained by the bridge. In other words, the piers would narrow the river, increasing its destructive velocity, potentially driving it over its banks where it would destroy all in its path. "I am finishing as fast as I can my new model of an iron bridge of one arch," Paine added. The society owed the people of Philadelphia a look at Paine's more inventive design.[7]

By late December, the new model was complete, and Paine and Hall had it brought to Franklin's house, just down the street from Carpenter's Hall. A few days after Christmas, Paine asked Hall to take the model to the Statehouse, where it would be displayed for the next few weeks. The state's assembly and its speaker reviewed the bridge, as did a parade of interested Philadelphians, among them "philosophers Mechanicks and even Taylors. Their sentiments of it was as different as their features," noted Hall. "The philosopher Said it wd Ad new lights to the world—The Mechanick it was Grand and Strong The Statesman with the merchant it wd be of great Utility— And the Taylor (for it is an absolute truth) remarked it Cut a pretty figure." It seemed that the proposed bridge would soon be a reality. The state assembly would incorporate the bridge company and "our Lilliputian Handyworke that is now 13 feet long . . . will be added . . . to the worlds present WONDERS."[8]

HALL'S OPTIMISM WAS MISPLACED. No bridge company would be incorporated for more than a decade. The politics were too contentious, the funds were still scarce, and the assembly was uncertain about the feasibility of so ambitious a structure. But Paine was not to be deterred. The praise his model had won convinced him that if he could only build a larger prototype, the virtues of the design would quiet any lingering doubt. Financing would follow. By March 1787, Paine had devised a plan.

Perhaps Benjamin Franklin had once again offered advice. The great man had returned from his diplomatic post in Paris in September 1785, just after Hall had arrived in the United States. During his near decade in the French capital, he had become one of the most celebrated figures in the city, admired in particular for his extraordinary electricity experiments—but also for his social appetite. No other American would have been better equipped to advise Paine on the prospects of finding support for a new invention in the capital of America's principal European ally. If it was Franklin who urged Paine to take up his bridge building in Paris, once again his counsel was sound. Although England was at the forefront of iron-bridge architecture, France had long been the world's most progressive country when it came to designing and building roads and bridges.

Early in the eighteenth century, King Louis XIV had founded the Corps des Ponts et Chaussées (Bridges and Roads Corps) and in 1747, this world's first engineering corps gave rise to the École des Ponts et Chaussées, the world's premier school of civil engineering and architecture. Given this expertise, Paine had concluded that were his newest model to receive the endorsement of the French and were they to afford him the opportunity to construct a larger prototype, the Americans would be persuaded, once and for all, that his was the soundest path across the Schuylkill. He was supported in this view by the Pennsylvania assembly, which appointed a committee of Robert Morris, George Clymer, assemblyman Thomas Fitzsimmons, and several others to review the French response. The assembly decided to delay any further decisions about the Schuylkill bridge until Paine returned to the United States the following winter.

ON APRIL 20, 1787, Paine and Hall met in Trenton to tend to some unfinished business. Paine still owed his model builder money, which he paid dutifully. Hall was happy to see his friend, who would be sailing for France in several days. The two spent the evening drinking and discussing affairs of state and the politics of bridge building. It was a fitting celebration of what had been, by any account, a fruitful partnership. Now, Paine told Hall, he would take the business of bridging the Schuylkill across the Atlantic. If the iron arch were to be sold to the people of Pennsylvania, it would need endorsement from the only country in the world with a corps of professional bridge builders. All that was left to do now was for the two to say their good-byes. "I then shook hands," Hall recalled, "and wishd him a good voyage and parted."[9]

8

American Architect

CARRYING LETTERS OF INTRODUCTION from Benjamin Franklin, Paine was welcomed in Paris. Franklin's French friends—particularly, Jean-Baptiste Le Roy, director of the King's Royal Laboratory in Passy—offered Paine all the trappings owed a close associate of France's great scientific hero. Le Roy introduced Paine to government and scientific officials, including Jean Perronet, a founder of the École des Ponts et Chaussées and the country's foremost bridge architect. Somehow Franklin thought Perronet had died and that Paine would have to settle for old drawings and models in place of the wisdom of the actual architect. But Franklin was mistaken and with Perronet's support, Paine received the expert audience he had come for.[1]

The French Royal Academy of Sciences would render judgment on the bridge. For a friend of Franklin, the academy could hardly have done otherwise. But the academy was interested in doing more than just humoring its American friend. Much like the rival imperial capital of London, Paris was perpetually constrained by inadequate bridges. As one contemporary described the principal bridges across the Seine:

> Two bridges designed for the communication with the court of Paris, are only miserable wooden bridges without decoration and almost without solidity: it is superlatively incommodious to find at the entrance of the one and the other, a gate through which two coaches cannot pass in front, without breaking, and to have a breadth upon these two bridges that is hardly sufficient for two coaches.[2]

Perronet rebuilt one of these bridges, the Pont de Neuilly. In place of a dilapidated wooden bridge dating from the reign of Henri IV, he constructed a stunning five-arch stone bridge, distinguished by its graceful narrow piers and its elliptical arches. The bridge, completed in 1774 and used until the middle of the twentieth century, placed Perronet at the leading edge of eighteenth-century bridge architecture.

He had built a bridge that elevated the merely functional into art worthy of an enlightened age. Its graceful low-slung arches and tapered piers gave it a lightness and elegance unusual for such a substantial masonry structure. Peronnet's contemporary, the British architect Sir William Chambers, described his contributions well when he wrote, "Materials in Architecture are like words in Phraseology . . . and may be so arranged as to excite contempt; yet when combined with Art, and expressed with energy, they actuate the mind with unbounded sway."[3]

For Perronet, the maxim extended beyond a structure's façade. By attending to the inner workings of his building, the architect opened new possibilities for aesthetic expression. He

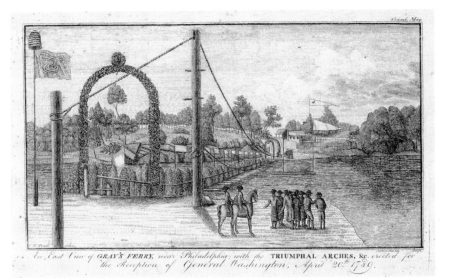

An East View of GRAY'S FERRY, near Philadelphia, with the TRIUMPHAL ARCHES, &c. erected for the Reception of General Washington, April 20.th 1789.

East view of Gray's Ferry floating bridge, prepared by Charles Willson Peale for the spring 1789 crossing of President George Washington. COURTESY OF THE PRINTS AMD PHOTOGRAPHS DIVISION, LIBRARY OF CONGRESS.

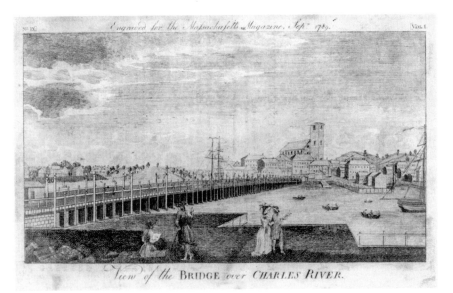

Engraved for the Massachusetts Magazine, Sep.r 1789.

View of the BRIDGE over CHARLES RIVER.

The Charles River Bridge, completed in 1786, was the first permanent bridge across the Charles. COURTESY OF THE PRINTS AND PHOTOGRAPHS DIVISION, LIBRARY OF CONGRESS.

London's Westminster Bridge undergoing repairs in 1749 after one of its supporting piers began to sink. CANALETTO (1697–1768), WESTMINSTER BRIDGE UNDER CONSTRUCTION, C. 1750; PEN AND INK. SUPPLIED BY ROYAL COLLECTION TRUST / COPYRIGHT © HM QUEEN ELIZABETH II 2012.

The Iron Bridge, erected near the Shropshire town of Coalbrookdale in 1779, was the world's first full-scale iron bridge. PHOTO: MELVIN GRAY.

Le Décintrement du pont de Neuilly, construit par Perronet (22 septembre 1772). Gravé par Prévost et de Longueil, d'après Eustache de Saint-Far.

The ceremonial 1771 striking of the temporary supports for Jean Perronet's new Pont de Neuilly masonry bridge before King Louis XV, his court, and hundreds of onlookers. AUTHOR'S COLLECTION.

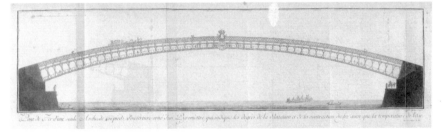

The thermometer at the top of Arnaud-Vincent de Montpetit's proposed 1783 iron bridge was intended to help measure the effects of changing temperature on the iron structure. COPYRIGHT © THE BRITISH LIBRARY BOARD; 8766.ee.1 (1)PLATE DE PONT DE FER.

The Sunderland Bridge over the River Wear was Britain's second major iron bridge when it opened in 1796. BY COURTESY OF THE TRUSTEES OF SIR JOHN SOANE'S MUSEUM. PHOTO: HUGH KELLY.

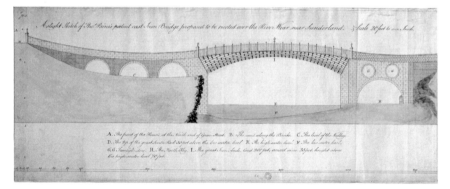

Design proposal for the Sunderland Bridge, attributed to Thomas Paine. Ink and watercolor, 1791. BY COURTESY OF THE TRUSTEES OF SIR JOHN SOANE'S MUSEUM. PHOTO: ARDON BAR-HAMA.

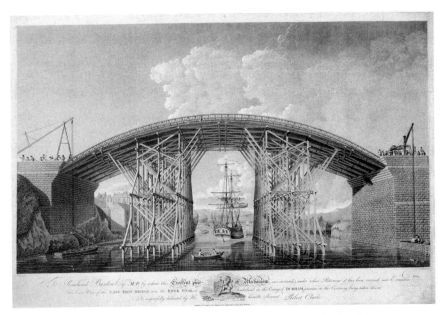

Unlike comparable stone bridges, the Sunderland Bridge was built without substantial interruption of river traffic. BY COURTESY OF THE TRUSTEES OF SIR JOHN SOANE'S MUSEUM. PHOTO: HUGH KELLY.

Thomas Telford and James Douglass's proposed Thames River iron bridge, presented to the Parliamentary Select Committee for the Improvement of the Port of London in 1801. COPYRIGHT © THE BRITISH LIBRARY BOARD; MAPS 3540 (14).

The 1802 Spanish Town iron bridge over the Rio Cobre in Jamaica remains the oldest and longest used iron bridge in the Western Hemisphere. COPYRIGHT © THE BRITISH LIBRARY BOARD; 1486.gg.1 PLATE 2.

Timothy Palmer's 1805 First Permanent Schuylkill River Bridge, with Owen Biddle's covering in the foreground. COURTESY OF THE PRINTS AND PHOTOGRAPHS DIVISION, LIBRARY OF CONGRESS.

The Swiss brothers, Hans Ulrich and Johannes Grubenmann, completed the covered Schaffhausen bridge over the Rhine in the late 1750s. WILLIAM PARS (1742–1782), *SCHAFFHAUSEN*, C. 1770; GRAPHITE, INK, AND WATERCOLOR. COPYRIGHT © TATE, LONDON 2015.

The Camelback Bridge across the Susquehanna River at Harrisburg, Pennsylvania, opened in 1817 and was among the country's longest covered bridges. COURTESY PENNSYLVANIA HISTORICAL AND MUSEUM COMMISSION, PENNSYLVANIA STATE ARCHIVES; IMAGE 8500.

The Lancaster–Schuylkill Bridge, more commonly known as the "Colossus," served travelers from 1813 until 1838, when it was destroyed by fire. EMMET COLLECTION, THE MIRIAM AND IRA D. WALLACH DIVISION OF ARTS, PRINTS AND PHOTOGRAPHS, THE NEW YORK PUBLIC LIBRARY, ASTOR, LENOX, AND TILDEN FOUNDATIONS.

America's first iron bridge opened in 1839 at Brownsville, Pennsylvania. COURTESY OF THE PENNSYLVANIA HISTORICAL AND MUSEUM COMMISSION, PENNSYLVANIA STATE ARCHIVES; IMAGE 5353.

was able, Perronet wrote, "to give us models of solid construction which without throwing away the elegant proportions presented by the monuments of antiquity will also approach the boldness and lightness of Gothic work." He could, in other words, build structures that at once exhibited the orderly symmetries of classical buildings and the soaring heights of the medieval cathedral.[4]

Paine never had any particular interest in his architectural inheritance. Whether it came from classical Rome or medieval Europe was of no consequence. What mattered were the basic tests of functionality; since, for Paine, function and form were united, a functional bridge would make for a visually striking bridge. The idea turned out to be compatible with Perronet's aesthetic. Although Perronet favored stone and timber bridges, the architectural impulses he represented fed a broader quest for more efficient use of materials and for structures that captured some of the wondrousness made possible by the medieval flying buttress. For some French bridge builders, the path to this new architecture lay through iron. With the appearance of the Iron Bridge, that path grew more attractive, and prospective bridge architects began promoting iron-bridge designs. Much like Paine, these French architects departed from the formal conventions of Western bridge architecture. In place of consecutive arches, they favored a single grand span; in place of high semicircular arches, they relied on arches fashioned from small segments of large circles; and their construction emphasized lightness by minimizing abutments and spandrels.

An iron bridge designed by Arnaud-Vincent de Montpetit, an inventor and painter, was typical. Much like Paine and John Hall, Montpetit was an obscure provincial who brought his appetite for invention to the city. Unlike Paine and Hall, he was born to a prominent family and possessed a substantial formal education. In the early 1750s, Montpetit moved to Paris, where he was recognized for a series of horological inventions, including an improved pendulum clock. He also dabbled in bridge architecture and designed a very early iron bridge for the Rhône that was never built. After a financial failure, he turned his attention to portrait painting and came to be known for portraits of the French king and for a method of painting on glass—a method that seemed to bring unusual luminosity to his depictions of members of the court.

His circumstances improving, Montpetit returned to his old vocation as inventor and began again to refine his iron-bridge designs. Although there is no evidence that Paine and Montpetit knew of each other's work, their designs were similar. Both relied on a low-slung iron arch, with a series of consecutive ribs bound together by iron cross members. In a 1783 petition to Louis XVI, Montpetit also made the case for an iron-arch bridge in terms almost identical to those Paine had been using. Although cascading ice was rare in France, river congestion and flooding were not, and Montpetit went to great lengths to demonstrate the many ways his sweeping arched bridge addressed these problems.

Beyond his petition, and an accompanying print depicting a four-hundred-foot version of the bridge, little came of Montpetit's design. Part of the problem may have been the basic tenets of his plan. Among the crucial benefits of iron-arch construction was its capacity to do what stone-arch bridges did but with less material, weight, and labor. Montpetit's design seems insensitive to this mandate. In addition to its rococo ornamentation, the bridge had two road decks, an enclosed one for pedestrians and another, above, for carriages. The bridge also featured a giant thermometer at the crown of its arch. This would allow users to gauge the effects of temperature on the structural integrity of the bridge. The idea was clever but it added to the complexity of the structure and highlighted the unknown effects of climate on iron architecture.

JUDGING FROM ITS RECEPTION, Paine's bridge suffered from no such defects. In summer of 1787, the Royal Academy appointed a special committee, with Le Roy as its chair, to review his and several other proposed iron bridges. The committee unanimously concluded Paine's was the "simplest, strongest, and lightest." But Paine needed more than mere plaudits. He needed to return to Philadelphia with a formal endorsement, something the committee seemed reluctant to issue. The problem, Paine learned, had to do with the structural principles behind the bridge. The review committee recognized that, with an astonishing lightness and elegance of form, the model could sustain weights proportionate to those the full-scale bridge would sustain. But why? What was it about Paine's configuration of iron bars that allowed them to bear such loads? Without a firm understanding of the structural principles behind Paine's iron bridge, the committee was initially reluctant to issue any definitive assessment.[5]

In late August, it nonetheless announced that, although its underlying structural principles remained unclear, Paine's was a design of supreme genius. Above all, it was unlike any other iron bridge, built or unbuilt, and yet it seemed entirely functional. Whether the full-scale version would equally distinguish itself, the committee could not say. There also remained a question about the effects of climate on iron. Would an iron bridge subjected to prolonged winter cold or summer sun maintain its structural integrity? And what would the effects of wind be on an iron arch of this scale? To address these and other questions, the committee recommended that Paine build a much larger prototype.[6]

Hoping the French endorsement would persuade the Pennsylvania assembly to push forward with the permanent Schuylkill bridge, Paine asked Franklin, who had been serving as president of Pennsylvania's Supreme Executive Council since the fall of 1785, to deliver the Royal Academy's report. Franklin did as Paine asked, but the assembly was slow to respond. Paine, meanwhile, carried on. To build a larger prototype, he needed to overcome new financial barriers. He had been able to pay for a thirteen-foot bridge, but anything much larger, especially the one-hundred-foot model Paine had begun imagining, would demand outside investment.[7]

In the fall of 1787, as he contemplated his next step, Paine returned to England to visit his ninety-one-year-old mother, who had been widowed the year before. During the visit, Paine reconsidered his prospects in France. Although the French had embraced his bridge, and although they were the world's best road and bridge builders, they had little experience with iron construction. Britain was the foremost iron-forging nation in the world. It was also filled with learned mechanics and wealthy patrons of science eager to promote the country's iron industry. Probably with Franklin's introduction, Paine began a correspondence with one of those patrons, the great Sir Joseph

Banks, president of England's Royal Society, Britain's most celebrated scientific institution. Banks encouraged Paine to present his model to the society, an opportunity Paine quickly seized. An endorsement from the Royal Society could draw the attention of Britain's ironworkers, the people who Paine assumed would have to build his new prototype. This would not be the work of toy-makers.[8]

FOR AN AMERICAN, and one with long-standing contempt for Britain's government, it was an especially exciting time to be in Britain. The American War fed a host of public and highly charged moves to reform the country and its empire. The abolitionists Granville Sharp and Thomas Clarkson, along with William Wilberforce, a young MP from Yorkshire, had begun a campaign to bring an end to the transatlantic slave trade—a campaign that would culminate in stunning Parliamentary debates over the morality of slavery. Corruption accusations leveled by Edmund Burke at Warren Hastings, the Governor General of India, would soon erupt into prolonged impeachment hearings at Westminster Hall. And throughout the British Isles, former supporters of the Americans were now hoping for more representative government at home.

Though the prospects for his bridge were bright, Paine could not resist the political fray. During the summer, Britain had begun preparing for war and Paine determined to intervene in the only way he knew how. Shortly after arriving in England, he published a lengthy pamphlet entitled *Prospects on the Rubicon.*

Here, just several years after the conclusion of the American War, the British government stared across the proverbial River Rubicon yet again. The cause of the impending conflict was potentially contagious unrest in the Netherlands. France had allied itself with republican partisans; Britain and Prussia with the Dutch stadtholder, the country's quasi-monarchical executive. By the summer of 1787, it appeared that Prussia would soon act against the republicans and their French allies, drawing Britain into yet another European war.

The crisis had implications for Paine's bridge. A Holland controlled by Prussia and Britain would likely curtail its merchants' abilities to do business in America. Similarly, France, still America's principal European ally, was itself staring into the fiscal abyss left by the American War. A new war with Britain might have been disastrous. If France and Holland, America's primary international lenders, were unable to extend the country credit, or if the war drove up the cost of existing debt, financing for Philadelphia's iron bridge would grow even more scarce.

These concerns would be meaningless to the British audience to whom Paine addressed *Prospects*. But a central preoccupation of Paine's, he concluded, would resonate. Britons were once again being drawn toward disaster by a government and its moneyed supporters for whom war had little to do with national security and everything to do with their own power:

It will always happen, that any rumor of war will be popular among a great number of people in London. There are thousands who live by it; it is their harvest; and the clamor which those people keep up in newspapers and conversations passes unsuspiciously for the voice of the people, and

it is not till after the mischief is done, that the deception
is discovered.

In the end those who paid dearest were the ordinary working
Britons on "whom the real burden of taxes fall."

Britain's reliance on war as statecraft, Paine warned,
created a vicious cycle of increasing debt and heavier taxes.
Britain's mixed constitution, with its democratic, aristocratic, and monarchical elements, seemed only to serve that
vicious cycle. With royal coffers, a monarch could buy off
the Commons and the Lords, the other governing constituen-
cies. This made both the royal patron and his clients hunger
for ever more tax revenue. And, once again, the state sated
its relentless appetite with the one reliable justification for
taxation: war. With no war, there would be no additional
taxation. And with the collapse of the fiscal-military matrix,
the government's ability to control opposition through graft
and corruption also disappeared. In bringing an end to Britain's pernicious politics, "wars, so fatal to the true interest"
of the people of Britain, "might exist no more, and . . . a long
and lasting peace might take place."[9]

Paine's enlightened idealism, resting on the proposition
that, as self-interested creatures, human beings had no inherent
appetite for war, was common enough among enlightenment
thinkers, particularly on the continent. Voltaire and Imman-
uel Kant propagated versions of the doctrine, as did the Marquis de Condorcet, a statistician and revolutionary who would
come to be among Paine's closest French associates.

But Paine's views were most compatible with those of
Thomas Jefferson, the American envoy to France. The two

shared a hostility to the British fiscal-military state, faith in the people's ability to govern themselves and to do so without resort to constant, ever-encumbering war, and devotion to the restorative effects of revolution. Although they knew of each other, and likely met in the early days of the Revolutionary War, until Paine arrived in Paris in 1787 they had had very little contact. But with Britain girding for war, they would begin a friendship that would last until late in Paine's life.

In 1784, Congress had appointed Jefferson to replace the retiring Franklin in Paris. French affection for the new nation and hatred of Great Britain meant that Americans found themselves welcome in the French capital. Dozens of them traveled there to trade, converse, and savor Parisian cultural life. Many would stay with Jefferson or dine at his residence. Paine was no different. Over dinner, he and Jefferson discussed Newtonian physics, the principles of iron-bridge design, and European affairs.

Like most of their countrymen, they also almost certainly discussed a recent popular uprising in western Massachusetts associated with Daniel Shays, a farmer and former Continental Army captain. The genesis of what came to be known as Shays's Rebellion lay in America's war-time fiscal crisis. The government of Massachusetts, struggling to repay its debts, had been raising taxes since the outbreak of the war. The taxes, which had to be paid in scarce hard money, proved a special burden to farmers, who were struggling to recover from war-time losses and wilting under an ever-heavier debt burden. Rather than provide relief for the farmers, the state encouraged the courts to begin seizing property from delinquent debtors. In an effort to stop the seizures, during the summer of 1786, Shays and hundreds of other farmers began shutting down the court system in the western part of the state and eventually attempted to seize the town of Springfield, home to an arsenal that had been used by the Continental Army. By February 1787, the state militia had brought the protest to an end, but subsequent elections yielded a state legislature more sympathetic to the farmers' plight.

In its broader effects, the uprising left creditors all the more apprehensive about public debt and sharpened calls for a radical revision of the Articles of Confederation. A stronger national government, many had come to believe, could control inflation, one of the problems that drove Massachusetts toward painful austerity in the first place. In the spring of 1787, about a month after Paine had left for France, those calls culminated in the convening of a national constitutional convention in Philadelphia. For Jefferson, a reconsideration of the Articles of Confederation had been long in coming. As chief envoy to France, he directly confronted the political consequences of a weak Congress. As long as European powers knew that Congress lacked the ability to levy taxes and had no real power to enforce trade agreements, Jefferson had little leverage abroad.

Paine too had long worried about the inadequacies of the Articles. In 1780, he had written of his desire to renew "a hint which I formerly threw out in the pamphlet 'Common Sense,' and which the several states will, sooner or later, see the convenience if not the necessity of adopting; which is, that of electing a Continental Convention, for the purpose of forming a Continental constitution, defining and describing the powers and authority of Congress." The Continental Congress had drafted the Articles of Confederation. This was antithetical to Paine's republican ideals. Government's powers should be established by the people, not the government itself. Without this kind of popular mandate, no government would be trusted to act for the common good.[10]

OVER THE NEXT three years, Paine's friendship with Jefferson would flower. Some of this grew out of the two men's shared scientific and mechanical interests. Although Jefferson lacked Franklin's influence in the European scientific community, he became a source of constant encouragement for Paine. More significant, though, was Jefferson's role in sustaining Paine's political activities.

Through the fall of 1787, after the appearance of *Prospects*, Paine tended to family business in Thetford and promoted his bridge around London. He returned to Paris that winter, but traveled back to England in the spring of 1788 to carry on with the bridge. Paine's return coincided with an important change in the American diplomatic corps. John Adams had been serving as American envoy to Britain since 1785. Adams had been directed to achieve a trade accord and to persuade Britain to withdraw its troops from the western reaches of American territory. But an intransigent and politically vulnerable British government gave little ground and a dejected Adams asked to be recalled. In the spring of 1788, he and his family returned to the United States. Congress was slow to send a successor, leaving Jefferson no reliable source of intelligence on British affairs. Working to advance American interests at a French court constantly eyeing its enemy across the English Channel, he needed some sense of Britain's intentions in Europe, particularly after the recent escalation of tensions surrounding the Netherlands.

Jefferson turned to his friend Paine, and for the next two years Paine would serve as an informal American agent in London. For Paine, as it became ever more apparent that he would not be returning to Philadelphia, the arrangement was ideal. It occupied him for the many weeks during which his architectural pursuits slowed.

9

An Architect and His Patrons

A S PAINE GREW ABSORBED with world affairs, his circle in England also expanded. Initially limited to old friends from his Lewes and Thetford days, it now included some of Britain's most influential men. In addition to providing information about affairs of state, Paine's new British friends proved indispensable supporters of his bridge building.

The most important such person was Edmund Burke. By the time he and Paine met in 1788, the fifty-nine-year-old Burke had been a Member of Parliament for more than twenty years. After half a decade as MP for the port city of Bristol, Burke had become MP for Malton, a small market town in North Yorkshire. It was a seat he held at the behest of the Marquess of Rockingham, his close friend and political patron, who had died in 1782. Rockingham's party, the so-called Rockingham Whigs, emerged as an organized opposition to the ministry of Lord North during the American War.

Reflecting much of his party's oppositional temperament, Burke had become a prominent voice of reform. He had opposed the American War on the grounds that it would irreparably damage Britain, whatever the outcome. He favored the extension of rights to persecuted minorities—particularly, Britain's disaffected Catholics. In the aftermath of the Gordon Riots, a violent anti-Catholic uprising that swept across London in June of 1780, he urged the government to respond with restraint. He opposed slavery and led a long and exhausting campaign to root out corruption in the British East India Company. In his defense of these causes, Burke came to be known as one of the great orators of his age. If he was not always able to influence an intransigent governing majority, he made his voice heard within and beyond Westminster.

BORN IN DUBLIN, the son of a prosperous Protestant attorney and a Catholic mother, Burke was raised a Protestant and educated by Quakers. After graduating from Trinity College, Dublin, he moved to London to pursue a career as writer and poet. Although his father pressured him to study law, the young Burke persisted and made his way among London's legions of scribblers. Since so much written for London's literary periodicals was anonymous, it is difficult to know the full scope of Burke's early work. But it almost certainly included magazine essays, sentimental verse, doggerel, and perhaps political pamphlets.

By his mid-twenties, Burke had begun accumulating a distinguished body of philosophical and historical work. His literary talents attracted the interest of prominent politicians, including the Marquess of Rockingham. In 1765, the year the King asked Rockingham to lead the government, Rockingham hired Burke as a personal secretary, a position that included a seat in Parliament. Rockingham remained in power for barely over a year, but Burke retained his seat after having led the administration's quest to repeal the controversial Stamp Act.

Politics may have been the initial basis for Burke's friendship with Paine. But the two men also shared an interest in the built environment. Early in his Parliamentary career, Burke had helped advance a bill to finance London's first substantial public office building, Somerset House. For the seat of so great an empire, the city had remarkably few distinguished public edifices. Many of its government offices were housed in ramshackle buildings around Whitehall and Westminster; similarly, its great learned societies—including the Royal Society, the Royal Academy of Arts, and the Society of Antiquaries— all convened in buildings of little distinction. It was time, Burke and his allies thought, for the capital to have a public building worthy of its imperial greatness.

The initial planning of Somerset House, named for the Duke of Somerset, whose sixteenth-century palace originally occupied the site of the new building, incited the kind of controversy public buildings always seem to stir up. Burke and others assailed the first design as crude and utilitarian, while others argued for economy of scale and style. When the architect responsible for this early design died, Burke's friend William Chambers, George III's chief architect, was chosen to propose an alternative. His building was really a grouping of connected neoclassical structures situated around a grand courtyard. In the front was the busy Strand, the thoroughfare linking the commercial City of London with the governmental City of Westminster. The rear of the building looked out upon a vast, stately terrace, perched high above the banks of the Thames and affording one of London's most stunning vistas.[1]

IN THE SPRING of 1788, Paine approached Burke with a letter of introduction from Henry Laurens, a South Carolinian, an old and devoted friend of Burke, and father of John Laurens, Paine's former diplomatic associate. During the American War, the older Laurens had been captured at sea, charged with treason, and imprisoned in the Tower of London. Burke was a vocal opponent of Laurens's treatment and helped negotiate his release. Once Paine met Burke at the latter's Soho residence, he became a regular in Burke's circle, which included the Duke of Portland, a former prime minister. The Duke invited Paine to Bulstrode, his Buckinghamshire estate. Paine would also be an occasional visitor to Burke's nearby country home at Beaconsfield.[2]

Early in the friendship, Paine had sought to separate himself from his radical past, telling Burke that he "had closed [his] political career with the establishment of the independence of America, and had no other business in France than to execute the orders of the government of Pennsylvania with the academy of sciences respecting the model of the Bridge." And although he had published *Prospects on the Rubicon*, this foray into global affairs was momentary: "The quiet field of science has more amusement to my mind than politics and I had rather erect the largest [iron] arch in the world than be the greatest emperor in it."[3]

BY THE SPRING of 1788, Paine had still heard nothing from authorities in Pennsylvania and had all but abandoned hopes of American financing for his bridge. Burke and the Duke of Portland gave him access to a new network of wealthy patrons, whom he now hoped would finance the prototype. Paine had also begun entertaining fantasies about bridging the Thames and using the profits from the sale of the bridge to finance a Schuylkill bridge.

The first step remained the construction of the new prototype. In addition to its architectural value, such a model would help Paine survive in Britain. He still had some money from the Continental Congress and was able to borrow against his American property, but funds were running low and his diplomatic work was strictly pro bono. The immediate solution, Paine believed, would be income from the display of his bridge model. London was awash in exhibitions of artistic and scientific curiosities. For a few shillings, Londoners could experience electricity shows, balloon ascents, and automata, humanlike machines that appeared to play chess, dance, draw, or play music. They could also view astonishing works of art and craft. In 1790, a gallery on Greek Street in Soho displayed Josiah Wedgwood's remarkable Portland Vase, a replica of the celebrated Roman original owned by the Duke of Portland.

Paine hoped to capitalize on this public appetite for science and art. While devising the scheme, he met an English-born Philadelphia merchant named Peter Whiteside, who promised to finance the new prototype on the condition that Paine acquire English, Scottish, and Irish patents. Paine had never intended to turn his bridge into proprietary technology, but Whiteside expected a return on his investment. Whiteside also recognized that Paine's iron arches could do more than support bridges and he insisted that these other applications get their

due. As Paine's patent made clear, his was a very broadly appli-
cable technology: "A Method . . . by Means of Which Construction, Arches, Vaulted Roofs, and Ceilings may be Erected
to the Extent of Several Hundred Feet Beyond what can be
Performed in the Present Practice of Architecture."

British patent law carried no authority in the United States
and Paine disclaimed any interest in a separate American patent. "With respect to the patents in England," he explained to
Jefferson, "it is my intention to dispose of them." For Paine,
the patents were means to an end: in securing Whiteside's
investment, they would allow him to build the bridge prototype upon which he had now staked his architectural future.[4]

However dismissive he may have been of his patents, they
showed remarkable architectural prescience. The age of cav-
ernous interior spaces of the sort that required iron ceiling
trusses would have to await the arrival of large-scale facto-
ries, exhibition halls, and the most triumphal expression of
industrial-age architecture, the railroad station. As Boulton
and Watt were transforming the British industrial landscape,
some of these sorts of buildings might have been imaginable.
But for an American, whose country remained an industrial
backwater, they were truly visionary.

MOST LIKELY through Burke's introduction, in August or
September of 1788, Paine met members of the Walker family, proprietors of one of Britain's largest and most profitable
ironworks. Founded in the 1740s by the late Samuel Walker,
the firm was now run by Walker's four sons, Samuel, Joseph,
Joshua, and Thomas. During the recently concluded Amer-

ican War, the firm's works in Rotherham, South Yorkshire, had profited mightily from the production of hundreds of tons of cannon. With little fanfare, the Walkers determined that Paine's bridge design was feasible and potentially profitable. They agreed to fabricate a new prototype and provide additional financing.

Paine never showed any qualms about the fact that the Walkers made their money supplying the British army in its war against his homeland. Nor did he ever acknowledge that his adventure into the age of iron would now be made possible by the very forces of war and state finance he so despised. In some sense, this silence reflected precisely the kind of hypocrisy Paine's critics would endlessly attribute to him. He was building bridges and writing political pamphlets not because he believed in what he was doing, but for base personal profit. Perhaps a more persuasive explanation, given Paine's ideals, is that he was willing to suspend scruples because he believed doing so would make possible the construction of an American bridge. And it was on the other side of the Atlantic that he expected his ideals to be fulfilled. Unless its people chose the path of revolution, Britain would always be a taxing, war-making nation. America, with its commercial population freed from the depredations of the British government, would be different.

By October 1788, Paine had traveled to Rotherham to begin work on a giant 250-foot bridge model. Shortly after his arrival, Francis Ferrand Foljambe, a local gentleman and Yorkshire MP, offered him his first British architectural commission. The River

Don ran in front of Foljambe's home, and could be crossed only by an "ill constructed" bridge. Foljambe's proposal that Paine build a bridge over the river was ideally timed. As the winter grew colder and darker, Paine had begun to rethink the size of his planned prototype. A smaller model could be built indoors and perhaps finished more quickly. It turned out that a smaller model was what Foljambe needed near his house. South Yorkshire would now have an iron bridge worthy of its growing iron industry. At ninety feet in length, the Foljambe bridge would be only slightly smaller than the famed Iron Bridge.

The fate of the Foljambe bridge is unknown, but it clearly fed a growing sense among the region's industrial and political elite that the American Paine was an architect worthy of their patronage. The project immediately attracted the attention of local dignitaries. The Rotherham works were just a few miles from Wentworth Woodhouse, the Marquess of Rockingham's vast 365-room manor, now occupied by Rockingham's nephew, heir, and political successor, the Fourth Earl Fitzwilliam. Shortly after Paine began working on Foljambe's bridge, Burke brought Fitzwilliam to see the marvel. The Earl was sufficiently impressed to invite Paine to spend time with him at Wentworth Woodhouse, an invitation Paine happily accepted.[5]

The year 1788 was a superb one for the Walker brothers. The firm issued £14,000 in dividends, some of which the Walkers agreed to use to finance yet another prototype. The new bridge would be displayed in London where, Paine hoped, it would attract paying visitors. Throughout 1789, Paine traveled back and forth between London, Paris, and Rotherham, alternately supervising the construction of the newest bridge, following what appeared to be the beginnings of revolution in France, and keeping up with British affairs in London. By the end of May, the following year, the Walkers had finished casting the parts of Paine's bridge and all 36½ tons of them were taken by ship to London.[6]

Paine would spend the next several months observing the construction process from a distance. He had hired an American, a Mr. Bull, to oversee the three carpenters and two laborers who would assemble the structure. The arrangement allowed Paine to keep a foot in the worlds of both architecture and politics. And it turned out that in the spring of 1790, the latter were particularly consuming.

REVOLUTION IN FRANCE had been under way for the better part of a year. The Estates General, a representative body that had not met since 1614, had been convened by the King, and the Third Estate, its quasi-representative branch, had decreed itself a new unitary National Assembly. The Bastille, a prison in central Paris, had been seized by partisans of the Third Estate. And the new self-declared National Assembly had abolished feudal privilege and seized church-owned property. The people of France, it seemed, had begun a wholesale reformation of government.

Although Paine watched these events with great anticipation, what most absorbed him were British affairs. Once again, it appeared, the government was preparing for war. After Spanish naval vessels captured two British merchant ships at Nootka Sound, on the remote west coast of present-day Vancouver Island, the government seemed bent on punishing its Spanish rival.

During the spring and summer of 1790, the crisis consumed Paine's political energies. He never acknowledged that his new patrons stood to profit from renewed war. Instead, he spoke of the crisis as another indication that Britain was in need of deep political reform. The government of Prime Minister William Pitt, he believed, had trumped up another excuse to raise taxes.[7]

By late summer, Spain had largely capitulated and the danger of war dimmed. Paine turned his attention back to his bridge. This was, it turned out, a necessary shift. In early August, Bull had fallen during a rainstorm, and he was now unable to work. Paine would have to oversee the final assembly of the new bridge. For the fifty-three-year-old Paine, the work proved exhausting but also exhilarating. "I am always discovering some new faculty in myself either good or bad," he wrote to Thomas Walker, "and I find I can look after workmen much better than I thought I could."

The construction site was carefully concealed behind a fence, but Paine reported many interested onlookers. He was as hopeful as he had ever been. A 110-foot iron-arch bridge, built entirely to his specifications, would now demonstrate once and for all the suitability of his design for the Schuylkill River.[8]

10

The Great Rupture

O N SEPTEMBER 15, 1790, the London *Public Advertiser* announced that the new iron bridge of Mr. Paine was being displayed and that it gave "infinite satisfaction." For a single shilling, visitors could now traverse the gentle arch above Lisson Green.

Concluding that fall was not the best season to attract more visitors, Paine closed the exhibition sometime in early October. In the spring, once the weather improved, the public would again be able to view the bridge. In the meantime, Paine would shelter his creation from the elements and make any necessary adjustments to his design. Unfortunately, by the end of November, whatever plans Paine had made for the winter months had been thrown into chaos. He had become embroiled in a shattering dispute with his friend Edmund Burke.[1]

PAINE'S FRIENDSHIP with Burke had shown signs of strain during the Nootka crisis. Although the two generally agreed about the belligerence of the Pitt administration, they had begun to disagree about the course of revolution in France. Paine remained steadfast in his belief that the French were simply doing what their American brethren had done. In place of an old, despotic mode of government, they were devising representative institutions and a just constitution. This did not necessarily mean deposing the King or destroying the monarchy. But it did mean reforms that would establish the Third Estate as the country's primary law-making body. As for Burke, although he had supported the Americans and had defended some of their claims against the British government, he had come to believe that events in France had little in common with revolution in the United States.

The two friends were at odds on other matters as well. One of these was the British government's treatment of religious minorities. Charles James Fox, a onetime political ally of Burke and former secretary of state, had been championing Parliamentary action against the Test and Corporation Acts, a series of old laws intended to exclude Catholics and other religious dissenters from public office. To the shock of many, Burke broke with Fox on the issue. As the Revolution in France progressed, Burke came to see dissent from the Church of England as a threat to the nation. Proponents of French-style radicalism, he feared, could conceal their true intentions—the wholesale destruction of Britain's governing order—behind a quest for religious liberty. Revolution in Britain would be a near certainty if such radicals were given access to government offices. To Paine, this position seemed perverse for a man with well-known Catholic sympathies, and by spring of 1790, the friendship had experienced a distinct chill.

Frustrated though he was by these disagreements, Paine had no sense of the full depth of his departure from Burke. While Paine was assembling his bridge, Burke had been fixated on events in France. Much like Paine, he had initially regarded the convening of the Estates General as the promising beginning of constitutional reform. But as events seemed to spiral out of control, and the people began to fill the streets of Paris, Burke's assessments turned very dark. In place of reasoned constitutional reform, the French appeared bent on mindless revolution. In defiance of any received constitutional traditions, the Third Estate had declared itself the nation's sole representative body, to the complete exclusion of the clergy and nobility, the other two estates. With the storming of the Bastille in July 1789, the nationalization of church property, the imposition of civil oaths on clergy, and the forcible removal of the King and Queen from Versailles to Paris, the consequences of this revolutionary act were plain to see: the wholesale rejection of the rule of law. The Americans had never, to Burke's mind, shown so much disregard for their institutional inheritance.

The problem, Burke had come to believe, was that France's revolutionaries had no sense of history's power. In their hostility to the standing order and its historical antecedents, Burke pronounced, France's revolutionaries were destroying the foundation of enlightened and civilized society. For what they mistook for that foundation—rights inherent in the natural order of things—were in fact human conventions, devised over time and assimilated by societies as they moved from barbarousness to civilization. In Burke's view, what ultimately made liberty possible, what allowed human beings to live by the rule of law rather than the yoke of tyranny, was not, as the French revolutionaries appeared to think, some sort of universal natural law. What secured the rule of law were the accumulated lessons of history, exemplified in that most noble of things, the English Constitution. The latter had emerged from a heroic and ongoing human struggle against its baser impulses. And the results of that struggle were all the conventions in law and society, all the manners and morals, all the institutions and traditions that allowed human beings to arrest the march of barbarousness. In their disregard for the constitutional rights of the nobility and clergy, in their refusal to see their king and queen as anything other than an ordinary man and woman, the French revolutionaries displayed a disregard for the very habits of mind that made civilized life possible.

In their place, they instituted a grotesque Hobbesian world where, as Burke so memorably put it, "all the decent drapery of life is to be rudely torn off. All the superadded ideas, furnished from the wardrobe of a moral imagination, which the heart owns, and the understanding ratifies, as necessary to cover the defects of our naked, shivering nature, and to raise it to dignity in our own estimation, are to be exploded as a ridiculous, absurd, and antiquated fashion."[2]

With the publication of his antirevolutionary _Reflections on the Revolution in France_ in November 1790, Burke would see his reputation soar. Within a few months of its appearance, 17,500 copies of _Reflections_ had been printed, an extraordinary number for a costly 356-page pamphlet. The pamphlet would go through eleven editions in its first year of publication. King George III praised the work and its author: "there is no Man who calls himself a Gentleman that must not think himself obliged to you."[3]

As REFLECTIONS TRANSFORMED BURKE into the conscience of establishment Britain, it would soon transform Paine into the conscience of radical Britain. In the fall of 1789, Paine had begun writing a history of the revolution in France but he abandoned the project to complete his bridge. When *Reflections* appeared the following November, Paine returned to his history, but what ultimately emerged in March 1791 was the first part of Paine's greatest political tract, *Rights of Man*. Far from the story of events in France, the pamphlet was a loud, fact-filled polemic. Much like *Common Sense*, it was a work of craft rather than high literary art, a fierce counter to Burke's reversions to history's mystical depths. It was also a deeply angry piece of writing.

Paine had read Burke's *Reflections* as a personal attack. With its contempt for the "swinish multitude," as Burke referred to the common masses, how could the democrat Paine have regarded it otherwise? For here was an attack on much more than the course of reform in France. Burke had directly challenged the very notion that ordinary people, people like Paine, were equipped to govern themselves, let alone anybody else. Human beings, Burke proclaimed, were not uniform in their character. As a result, they were not uniform in their capacities.

> The occupation of a hair-dresser, or of a working tallow-chandler, cannot be a matter of honour to any person—to say nothing of a number of other more servile employments. Such descriptions of men ought not to suffer oppression from the state; but the state suffers oppression, if such as they, either individually or collectively, are permitted to rule.[4]

How could a former stay-maker, turned revolutionary, now in pursuit of the world's highest ideals as architect, not be offended by these words, especially coming from a close friend and patron? "From the part Mr. Burke took in the American Revolution, it was natural that I should consider him a friend," Paine explained, "and as our acquaintance commenced on that ground, it would have been more agreeable to me to have had cause to continue in that opinion, than to change it." But Burke left Paine no choice. Not only would he have to abandon any pretense of friendship, he would have to stand up for the values that had motivated his fellow American revolutionaries and that now inspired the revolutionaries of France. Among the most fundamental of these was the idea that government by the people was the only truly just government:

> Mr. Burke will not, I presume, deny . . . that governments arise, either *out* of the people, or *over* the people. The English government is one of those which arose out of a conquest, and not out of a society, and consequently it arose over the people; and though it has been much modified from the opportunity of circumstance . . . the country has never yet regenerated itself.[5]

In abolishing the seigneurial rights of the French aristocracy and the Catholic Church, and in refashioning the Third Estate into an elected and representative National Assembly, France was bringing an end to government above and beyond the governed. Indeed, what it was doing was precisely what Britain ought to do: abandon the charade of so-called mixed British-style constitutionalism. Much as he had done in *Common Sense*, Paine proclaimed this mixed constitution to be nothing more than "an imperfect every-thing, cementing and soldering the discordant parts together by corruption." With no accountability and no ultimate authority, the system created a government of dysfunction:

> In mixed Governments there is no responsibility: the parts cover each other till responsibility is lost; and the corruption which moves the machine, contrives at the same time its own escape. When it is laid down as a maxim, that *a King can do no wrong*, it places him in a state of similar security with that of idiots and persons insane, and responsibility is out of the question with respect to himself. It then descends upon the [Prime] minister, who shelters himself under a majority in Parliament, which, by places and pensions, and corruption, he can always command; and that majority justifies itself by the same authority with which it protects the Minister. In this rotatory motion, responsibility is thrown off from the parts, and from the whole.[6]

The only Englishmen served by England's mixed constitution, in Paine's view, were the King and his ministers. Meanwhile, everybody else paid to keep them in their foppery and finery.

Burke's defense of this archaic and destructive order was all the more puzzling, given his former sympathies for the Americans. As he knew only too well, they had been governing themselves for some time without king and court. How could he now believe that France, in creating representative government, was marching headlong into the historical abyss? The answer was one Paine believed Burke knew well but refused to acknowledge:

It is easy to conceive that a band of interested men, such as Placemen, Pensioners, Lords of the bed-chamber, Lords of the kitchen, Lords of the necessary-house, and the Lord knows what besides, can find as many reasons for monarchy as their salaries, paid at the expense of the country, amount to; but if I ask the farmer, the manufacturer, the merchant, the tradesman and down through all occupations of life, through the common labourer, what service monarchy is to him? He can give me no answer.

The reason was simply that ordinary people, whatever their country, know little of the government of ministers and kings. They know government as the protector of property rights, the mediator in civil disputes, and the agent of internal improvement. This government, even in England, was not the government of monarchs and political appointees, or placemen, but the republican branch of government. Paine referred to this kind of local, personal, and benevolent government as civil government, and in his view all civil government was republican government. This was even true in monarchical Britain. After all, "that part of the government of England which begins with the office of constable, and proceeds through the department of magistrate, quarter-session, and [courts of] general assize, including trial by jury, is republican government. Nothing of monarchy appears in any part of it."

This government, not the King in Parliament, was what ordinary Britons knew as government. And yet, they found themselves paying for something that did very little for them. In effect, "the Nation is left to govern itself, and does govern itself by magistrates and juries, almost at its own charge, on republican principles," while its taxes go to the monarch and his favorites.[7]

HAD IT NOT BEEN for *Rights of Man*, Burke's pamphlet might stand as eighteenth-century Britain's best-selling political tract. But Paine's essay was a spectacular publishing success. By the end of May 1791, it had gone through six English-language editions, with some 50,000 copies sold. Countless cheap pirated editions also appeared, as did German, French, and Dutch translations. In some ways, the content of the pamphlet was less momentous than its popularity. Over the three years after its initial appearance, between 100,000 and 200,000 copies were printed.

For Britain's Tory establishment, who now found new admiration for the opposition Whig Edmund Burke, the popularity of Paine's essay validated all that Burke said. The crude ideas of France's revolutionaries had begun to seduce the ignorant rabble of Britain and if Paine were allowed to continue his new campaign, the cancer of revolution would surely reach British shores. Thrilled by this response, Paine prepared a second part of *Rights of Man*, a deliberate and successful act of provocation. Part II was published in February 1792. In May the government initiated a seditious libel case against Paine, largely on the grounds that *Rights of Man* had not confined its appeal to the "judicious reader," but had instead attracted those "whose minds cannot be supposed to be conversant with subjects of this sort . . . the ignorant, the credulous, the desperate."[8] Paine, it seemed, had violated a principal tenet of public discourse. He did not write for the schooled intellectual or knowing statesman. He wrote for ordinary working Britons, and his astounding popularity was frightening confirmation that few of them accepted Burke's antirevolutionary ideas. As one correspondent of Prime Minister Pitt reported, the country is "covered with thousands of Pittmen, Keelmen, Waggonmen and other laboring men, hardy fellows strongly impressed with the new doctrine of equality, and at present composed of such combustible matter that the least spark will set them in a blaze."[9]

The cause of all this clamoring democratic madness was the American Paine. The conservative press responded with fierce denunciations and caricature. "It is not surprising," observed a letter writer in the London *Public Advertiser,* "in this age of novelties, to see an illiterate [excise collector], who had deserted his own country, return again to it, and enter the lists of controversy with a man of the first genius and learning of our times; nor even, to see him set up *as a teacher* in politics and government to this enlightened country, from which he absented himself, but to betray it." Even some who might have been allies now denounced Paine. The Reverend Christopher Wyvill, a political reformer and supporter of the radical nonconformist minister Richard Price, whose sermons in defense of the French Revolution helped drive Burke to write *Reflections,* announced with alarm that "the avowed purpose of [Paine] is not to reform or amend the system of our Government, but to overturn and destroy it."[10]

11

The Specter of Paine

A LTHOUGH PAINE WAS CONSUMED BY his ideological clash with Burke, financial necessity demanded that he attend to his bridge. He had returned most of his profits from *Rights of Man* to the printers as a means of keeping the price low. Peter Whiteside and the Walker brothers expected returns on their investment, and his personal reserves much diminished, Paine needed the income from the bridge exhibit.

By April 1791, Paine again welcomed visitors to Lisson Green. In an inept attempt to attract more of them, he and Whiteside took out newspaper ads announcing that in six weeks the bridge would leave London for Paris. A rush of new visitors never materialized and the ad provided fodder for the conservative London press, which lambasted Paine for using British money to finance a bridge that would ultimately be erected in its enemy's capital city. "The King of France," noted the *General Evening Post*,

> has purchased Mr. Paine's Iron Bridge, for which he paid
> him a considerable sum. The *Patriot* has got his reward;

this, together with the shillings he has received in this
country for seeing the bridge, will enable him to make
great improvements on his plantation in the Jersey's [sic],
the present of the Congress for his political labours in
America.—*Patriotism is a lucrative trade.*[1]

The bridge remained in place over the course of the summer.
But the enterprise was now in a free fall. Paine's own financial
fortunes continued to decline and his lease on Lisson Green was
about to expire. By the end of August, he was asking the Walkers to "take some steps in this [bridge] business as the time has
expired for which the ground was hired and I cannot possibly
charge myself any longer with the care of the concern."

Through the early fall, the bridge remained but by the end
of the year it had been dismantled and shipped back to Rotherham. Those who would see Paine's bridge had, by now, seen it.
And while much of the response was colored by Paine's political
activities, the model itself was no triumph. The bridge rested
between two small wooden abutments and over the course of
the seasons these yielded, causing the structure to sag. Paine was
somewhat dejected, but not despondent. To Hall, he wrote that
although the Lisson Green model had problems, they were not
irreversible. The first Rotherham prototype, which was probably the bridge erected on Foljambe's estate, had been "erected
between two steel furnaces, which supported it firmly; it contained not quite three tons of iron . . . was loaded with six tons
of iron, which remained upon it a twelve month," and there was
no sign of failure. The design, Paine remained convinced, was
sound. Whether he would be able to continue its refinement
was another matter.[2]

UNTIL JUNE 1791, PAINE and most other observers had assumed that as long as Louis XVI and his family remained in Paris, a new National Assembly would enact reform with the King's consent. There would be no need to abolish the monarchy. But if the monarch fled or otherwise disavowed the revolution, the path toward government without a king would be unavoidable. Radicals would point to royal duplicity as justification for the creation of a French republic.

In the end, this is precisely what happened. On June 21, the King and his family fled Paris, disguised as common travelers. They nearly managed to escape to the Dutch dominions of Leopold II, the Austrian emperor and brother of the French queen, Marie Antoinette. However, they were apprehended a little over thirty miles from the border, at the town of Varennes. For French republicans, the King's flight was proof that the Revolution could no longer protect a deceitful Bourbon monarchy.

Paine, who was in Paris at the time of the King's flight, interpreted events similarly and immediately joined the republican cause. He helped found the new Société des Republicains, whose members included the journalist and politician Jacques-Pierre Brissot de Warville; Achille François du Châtelet, a young aristocrat who served as Paine's translator; and the forty-eight-year-old statistician the Marquis de Condorcet. Paine was also involved in founding the club's short-lived journal, *Le Républicain*, a role well justified, he explained, by his citizenship in the United States, "a land that recognizes no majesty but that of the people, no government except that of its own representatives, and no sovereignty except that of the laws."[3]

When Paine returned to England later that summer, he

discovered that his fealty to French republicanism had further
alarmed the Pitt government. Rather than tolerate an embold-
ened Paine, the government undertook a carefully orchestrated
smear campaign, propagating rumors about financial malfea-
sance, commissioning a malicious biography by a Scottish lawyer writing under the pseudonym of Francis Oldys, sponsoring
new anti-Paine royalist associations, and covertly encouraging
a flood of anti-Paine pamphlets.

During the fall of 1791, as government pressure against him
grew, Paine sought refuge at the London home of an old friend
from Lewes, a bookseller named Thomas "Clio" Rickman.
Encouraged by his many like-minded friends, Paine responded
with Part II of *Rights of Man*. "As revolutions have begun," it
proclaimed, "other revolutions will follow." The message could
not have been plainer: the tide of revolution had turned, and
whatever Edmund Burke may have wished to believe, Britain
would soon follow France on the path to republican government:

> Never did so great an opportunity offer itself to England,
> and to all Europe, as is produced by the two Revolutions
> of America and France. By the former, freedom has a
> national champion in the Western world; and by the latter,
> in Europe. When another nation shall join France, despo-
> tism and bad government will scarcely dare to appear. To
> use a trite expression, the iron is becoming hot all over
> Europe. The insulted German and the enslaved Spaniard,
> the Russ and the Pole, are beginning to think. The present
> age will hereafter merit to be called the Age of reason,
> and the present generation will appear to the future as the
> Adam of a new world.

Revolution was now as much a fact of the Old World as the New. "What pace the political summer may keep with the natural," Paine concluded, "no human foresight can determine. It is, however, not difficult to perceive that the spring is begun."[4]

And all of this was the result of a simple truth: governments of war and taxes, of meaningless pomp and splendor, were giving way to governments by the people. In addition to openly prophesying revolution in Britain, *Rights of Man* put forward a radical plan to revamp the British system of taxation. Government by the people, liberated from endless dynastic and imperial war, would bring massive tax cuts, especially for poor and working Britons. The remaining revenue could be applied to a new system of social security for the poor, particularly the legions of British military veterans. In a letter he wrote not long after the appearance of Part II, Paine summarized his plan:

> The work shows . . . that the taxes [of Britain] now existing may be reduced at least six million, that taxes may be entirely taken off from the poor, who are computed at one third of the nation; and that taxes on the other two thirds may be considerably reduced; that the aged poor may be comfortably provided for, and the children of poor families properly educated; that fifteen thousand soldiers, and the same number of sailors, may be allowed three shillings per week during life out of the surplus taxes; and also that a proportionate allowance be made to the officers, and the pay of the remaining soldiers and sailors be raised.[5]

Paine knew well that this call for revolution in state finance would only further incite his growing list of enemies. He also knew that the government would use his latest publication as yet another justification to attack his character. It would point out that a man who once served its revenue system now identified that system as a source of social injustice. Paine would counter these claims with a propaganda campaign of his own. He wrote old associates in Lewes, reminding them of "the exceeding candor, and even tenderness, with which" he carried out his gauging. "The name of *Thomas Paine* is not to be found in the records of the Lewes' justices, in any one act of contention with, or severity of any kind whatever toward, the persons whom he surveyed, either in the town, or in the country."[6]

Paine's efforts did little to quiet the alarm produced by his latest pamphlet. Part II of *Rights of Man* sold as well as Part I. In addition to the tens of thousands of legally printed copies, thousands of copies of cheap pirated editions appeared throughout England, Wales, Scotland, and the United States. In Ireland, more than anyplace else in the British Isles, the pamphlet struck a chord. Long oppressed Catholics, dissenting Protestants, and republicans savored Paine's no-holds attack on Britain's constitution. For the government, Paine's popularity in Ireland constituted a frightening threat to national security. Despite its deep and conservative Catholicism, no population in Britain was more susceptible to French radicalism than the legions of alienated Irish Catholic peasants. For many of them, the Revolution's hostility to monarchical government eclipsed its deep anticlericalism. The danger of radicalism in Ireland grew all the more acute with the emergence in 1791 of the United Irishmen, a reform party composed of disaffected Irish Presbyterians and middle-class Catholics. One of the group's founders and its most prominent spokesman, a Protestant lawyer from Dublin named

Theobald Wolfe Tone, would come to know Paine as a man who "drinks like a fish," but who nonetheless "has done wonders for the cause of liberty, both in America and Europe." [7]

Initially Tone and the United Irishmen favored incremental reform more in the vein of the reformist Burke and his Whig allies than the radical Paine. Nonetheless, Britain's government saw signs of Painite republicanism in the group's calls for greater representation and Catholic voting rights. The possibility that such radical sentiment would spread throughout the country's embittered sectarian factions, encouraging revolution, was something the Pitt administration could not tolerate. The loss of American colonies to republicanism had already driven Lord North and his ministry from office. Pitt could not allow Irish radicalism to do the same to his government. And if this meant a heavy-handed response to Paine's latest revolutionary publication, so be it.

IN MAY 1792, the government issued a summons to Paine's publisher, J. S. Jordan, to appear before the King's Bench on charges of seditious libel. Jordan pled guilty and surrendered materials relating to Paine. Several days later, the government issued Paine a similar summons. Paine considered the legal action a conspiracy engineered by Burke to discredit the writer who had defeated him in the era's great ideological contest. [8]

Even though a conviction would mean imprisonment, Paine saw in his persecution signs of victory. He had succeeded in shaking conservative England and had every expectation of furthering his cause through a public trial. But the authorities were a step ahead of him. They delayed any trial until December, hoping that anti-Paine demonstrations, with their loud loyalist proclamations and executions of Paine in effigy, would force Paine to confess his crimes or leave the country.

The strategy worked.

On the evening of September 13, 1792, under cover of darkness, Paine fled for Dover, intending to sail the next day for Calais. The journey proved harrowing. Paine was detained by Customs officers, who forced him to surrender personal papers, including the page proofs for a third part of *Rights of Man*. The following morning Paine was released and, passing through a gauntlet of shouting, spitting local loyalists, he boarded a boat for France. He would spend the next decade there and never again return to England. The Lisson Green bridge, as far as he knew, was now a disassembled heap, rusting away at the Walker Brothers' foundry.

In December of 1792, the Pitt government began court proceedings against Paine. Given Paine's absence, the trial was obviously for show, but was undertaken with urgency, as antigovernment unrest erupted throughout the country, much of it stoked by rising food prices, Painite republicanism, and the prospects of war against revolutionary France. Although Paine's defense lawyer was Thomas Erskine, attorney general to the Prince of Wales, and among the nation's best lawyers, the carefully selected jury found Paine guilty. Far from quashing Painite sympathies, however, the verdict only seemed to excite Paine's supporters, thousands of whom rallied in the streets of London as Erskine's carriage made its way from Guildhall at the trial's conclusion.

DESPITE ALL THE DOMESTIC tumult, industrial Britain continued its forward march. The nation's second major iron span
was opened across the River Wear in 1796 at the northeastern town of Sunderland. Although the 236-foot bridge never
achieved the iconic status of the Iron Bridge, it was a much more
ambitious and majestic structure. For a region whose coal had
been heating London and fueling the new steam-engine boom,
the bridge was a triumph, a testament to northern England's
capacity for industrial development.

Capturing this spirit, although with a sharper note of aspiration than its authors probably intended, the foundation stone
of the bridge's abutment, which had been laid on September
24, 1793, paid tribute to the bridge's creator, a Member of
Parliament for the County Durham and a Tory ally of Prime
Minister Pitt:

> At that time when the mad fury of French citizens, dic-
> tating acts of extreme depravity, disturbed the peace of
> Europe with iron war, Rowland Burdon, Esq. aiming at
> worthier purposes, hath resolved to join the steep and
> craggy shores of the river Wear with an iron bridge.[9]

To a degree, the inscription was accurate. The Sunderland
Bridge was the work of a northern gentleman, one whose
family estate at Castle Eden stood as a monument to con-
servatism and, in this case, its mercantile roots. Rowland
Burdon's father had been a founder of the Exchange Bank
in Newcastle and Burdon himself had extended the bank's
business to the southeast, facilitating the growth of the

Sunderland port and financing the construction of the new bridge.

The opening of the bridge in August 1796, reported one London paper, drew a who's who of establishment Britain, featuring "a grand Masonic procession, attended by the Commissioners of the River Wear, Magistrates, clergy, Officers of the Navy and Army, and the Loyal Sunderland Volunteers." These dignitaries had come together to commemorate a great improvement in local life. The small colliers and coal barges feeding the North Sea coal trade could now travel uninhibited up and down the river while those traveling overland between Sunderland and Newcastle freely moved across the bridge. It was a fitting tribute to a British industrial revolution untouched by a French political one. [10]

Beneath all the righteous commemoration lay deep social tensions. Improving navigation had been, for Burdon and his constituents, a matter of vital economic importance. The region's keelmen—the small barge operators who hauled coal downriver to oceangoing ships, anchored in deeper waters—had periodically blockaded the Wear and Tyne Rivers during the course of the eighteenth century in what were among Britain's earliest labor strikes. A bridge that carried traffic above the river without further inhibiting boat traffic below could do much to mitigate the effectiveness of the keelmen's strikes.[11]

From the vantage of its builders, then, the gently curving single-arch Sunderland Bridge was a monument to capitalism and counterrevolution. Thomas Paine, one might well conclude, was the last person Burdon and his associates would have had in mind as they erected their iron bridge. And yet, observers and chroniclers of the bridge regarded it as an architectural wonder inspired by the vision of Thomas Paine.

BURDON DEVELOPED an interest in architecture in the mid-1770s during a grand tour of Italy. He and a group of fellow students traveled around the country studying its architectural monuments, particularly those built by the sixteenth-century architect Andrea Palladio. Among the lifelong friends Burdon made was John Soane, a young English architecture student who as professor at the Royal Academy would shape nineteenth-century Britain's first generation of architects.

As he was planning the bridge across the Wear, Burdon had consulted his old friend Soane, who was then working as architect and surveyor for the Bank of England in London. Burdon was most interested in Soane's views on iron construction and sent him a drawing to evaluate. The drawing was entitled *A Slight Sketch of Thos. Paine's Patent Cast Iron Bridge Proposed to Be Erected over the River Wear near Sunderland.* The only surviving image of Paine's bridge was most likely made by the Walker brothers to advance a bid to build Burdon's bridge. Soane made two copies of the drawing in November 1791. These remain in his papers along with a careful report on Paine's Lisson Green bridge.

Soane's assessment of Paine's bridge was guarded. "On reviewing the progress and improvement made in various arts and sciences," he noted, "I believe it will be found that most of them have been made by ingenious men not brought up to or engaged in that particular science, branch of business or manufactory." So it was that "perhaps we have seen a woman's stay

maker outstrip [Britain's foremost bridge architects] in constructing bridges of Iron . . . without any farther pretensions
to a knowledge of Architecture than what a moderate share of
common sense may afford."

But a bridge across the Wear required much more than
common sense. Even supposing an iron arch would function
as Paine assumed it would, "there are many other things to
be known . . . such as lying foundations, building piers, constructing centres [or temporary supporting scaffolds] for turn-
ing the arches upon & c." Above all, though, the integrity of
the iron-arch bridge would be an aesthetic matter. "Properly
proportioning one part to another so as to give the whole
structure the greatest strength and elegance" was, in the end,
something only the most "judicious and practical architects"
could ensure. Paine was very clearly not one of these.[12]

Encouraged by Thomas Wilson, a local Sunderland architect, Burdon nonetheless settled on an iron bridge very similar
to Paine's. To construct the bridge, he turned to the Walkers,
now Britain's foremost iron-bridge fabricators. They had urged
him to build with iron from the start and, to promote the idea,
may even have re-erected Paine's Lisson Green bridge at Castle
Eden, the Burdon family estate in County Durham. What the
Walkers made for Burdon was different in its specifics from
what they had made for Paine. Instead of a series of parallel
iron arches composed of smaller, curved bars, with a road deck
resting directly on the arches, the Sunderland Bridge consisted
of a series of arches, as well as a detached deck supported
by circular iron spandrels. But the fundamental form of the
bridge, a shallow iron arch composed of smaller load-bearing
iron members, was still Paine's.

Such, in any case, was the common perspective; however minimal Paine's actual role in the creation of the bridge, it stood in the minds of contemporaries as a tribute to the architectural genius from America. "The greatest object of curiosity in Sunderland is its iron bridge thrown across the river Wear, forming an arch so lofty as to allow large ships to pass under it," wrote the traveler and geographer Reverend Richard Warner. "Tom Paine . . . was the original inventor of these extraordinary structures," surely among "the grandest specimens perhaps, of the powers of modern art." In fact, according to his lecture notes, prepared in 1811, after he had become professor of architecture at the Royal Academy, John Soane had concluded that the iron bridge "was introduced to England by Tom Payne." Some claimed that the Sunderland Bridge was fabricated from the remains of Paine's Lisson Green bridge. The 1847 edition of Edward Cresy's *Encyclopedia of Civil Engineering* noted, for instance, that "Sunderland Iron Bridge, over the Wear, was formed in part out of another contrived by the celebrated Thomas Payne, which was put up at the Yorkshire Stingo, Lisson Grove, and afterward carried back to Rotheram."[13]

In 1859, as a refurbished Sunderland Bridge was about to be opened, planning for commemorative events was disrupted by scandal. The board overseeing the project made plans for a tribute to its original architect but it was divided over who most deserved the honor. Burdon's son, the Reverend John Burdon, offended by the possibility that his father would be denied his due "while Paine . . . is promoted to the place of honour," attacked the proceedings. In a lengthy published letter, he claimed that his father was a victim of the crude and narrow thinking of modern professionals: "I cannot help fancying a good deal of the doubt which has been cast upon this matter has originated in . . . a certain vague notion, abroad now-a-days, that a mere country gentleman is not to be expected to possess any great qualities of mind or character; and that, therefore, there is great antecedent improbability that one of that class should have hit upon . . . anything so remarkable as the invention of the Bridge."

Burdon's pleas had little impact. "At the Iron and Steel Institute, at Liverpool," reported the *Sunderland Daily Echo and Shipping Gazette* in 1879, "Mr. J. A. Picton, F.S.A., in the course of his paper on 'Iron and Steel as Constructive Materials,' on Thursday last, is reported to have said: 'The boldest [use of these materials] . . . was the cast iron bridge over the Wear [at] Sunderland, which was designed by the celebrated Thomas Paine and was opened in 1796."[14]

PAINE LEARNED OF the Sunderland Bridge sometime in the mid-1790s. Assuming it was a copy of his Lisson Green model, he asked a Paris friend and former Member of Parliament, Robert Smith, to help him obtain compensation. Smith wrote an old colleague, the Whig Ralph Milbanke, Rowland Burdon's fellow MP from County Durham. Although he had been an early supporter of the bridge, Milbanke had little to do with its design and construction and was able to offer Smith little more than acknowledgment of the bridge builders' debt to Paine:

With respect to the iron bridge over the river Wear at Sunderland, . . . I have good grounds for saying that the first idea was suggested by Mr. Paine's bridge. . . . With respect . . . to any gratuity to Mr. Paine, though ever so desirous of rewarding the labors of an ingenious man, I do not feel how . . . I have it in my power . . . but if you can point out any mode according to which it would be in my power to be instrumental in procuring him any compensation for the advantages the public may have derived from his ingenious model, from which certainly the outline of the bridge at Sunderland was taken, be assured it will afford me very great satisfaction.[15]

Under different circumstances, Paine might have taken some encouragement from Milbanke's words. He did, after all, have a patent for his bridge. Surely the younger Paine would have done so, or at the very least a Paine not wanted for seditious libel.

12

Citizen Paine

SHORTLY BEFORE FLEEING ENGLAND, Paine had learned that a provision in France's new constitution allowed foreigners to gain French citizenship. In Paine's case, the provision was meaningful because it also allowed him to be elected in absentia to represent the *département* of Pas-de-Calais in a national constitutional convention. The new body was convened in September 1792, amid the growth of antimonarchical anger and an expanding French war against Austria and its continental allies. For a man who had spent most of his adult life decrying the constitutional inadequacies of monarchical government, the opportunity to do what he did very little of in the United States—to help fashion a republic's fundamental governing principles—would seem a godsend. But as Paine soon discovered, constitution making in revolutionary France involved a series of problems he had thought little about.

The most momentous of these was the question of what was to become of Louis XVI and the rest of his Bourbon lineage. In the aftermath of the King's flight, and with growing evidence

that the royal family had covertly encouraged a loyalist counterrevolution, even moderates were flocking to the republican
side. In Paris, the thronging crowds had long since abandoned
hope of a constitutional monarchy. The *sans-culottes*, a diverse
collection of Parisian artisans, mechanics, shopkeepers, and
petty merchants, named for their choice of attire—the long
trousers of the artisan rather than the short *culottes*, or knee
breeches, of the social elite—had prepared the way for the
First Republic. In early August 1792, they joined forces with
national guard units to drive the King and his family from the
Tuileries Palace, an action that effectively deposed the monarch but also resulted in the brutal massacre of several hundred
Royal Swiss guards.

When the National Convention of 750 elected deputies
assembled the next month, on September 20, just a few days
after Paine's arrival in Paris, among its very first acts was to
affirm the abolition of the French monarchy. One year later,
the revolutionaries instituted a new calendar that established
September 22, 1792, as the first day of the first year of the
new French republic. To the horror of Burke and other conser-
vatives, constitutional reform in France involved much more
than the creation of a republic. It also involved starting history
anew.

Within days of the new republic's birth, the Convention's
deputies began pushing for the destruction of remaining ves-
tiges of monarchical authority, among them the judiciary,
established under the previous constitution, which had been
endorsed by the King. Georges-Jacques Danton, a now famous
young leader of the radical Parisian Jacobin club, led the Convention's assault against the judiciary, arguing that the only

way to establish its fundamental loyalty to the new republic was through the creation of a new corps of elected judges. In his first speech before the Convention, Paine countered that elected judges could never be impartial. In a sign of things to come, he was ignored. By late November, a debate in the Convention over whether to try the King for conspiracy, treason, and a long list of other crimes came to a sudden end. A sheaf of incriminating papers discovered in the vacant Tuileries Palace left no doubt about Louis XVI's guilt. The Convention would now have to decide whether or not the King—whom Paine had begun referring to by his family name, Louis Capet—would suffer the same fate as any other traitor to the state: execution. Paine and other deputies urged caution, fearing that execution would strengthen counterrevolutionary forces at home and abroad. It could also alienate French allies, particularly the new United States, a country that very likely owed its existence to Louis Capet. Meanwhile, the restive crowds of Paris were calling for royal blood.

As Louis's trial began in December, Paine pleaded with his fellow deputies to spare the former monarch's life, noting that the increasingly influential leader of the radical Jacobin Club of Paris, Maximilien Robespierre himself, had favored the abolition of capital punishment. Paine proposed, instead, that Capet be imprisoned until he could be banished to the United States. There, "far removed from the miseries and crimes of loyalty, he may learn, from the constant aspect of public prosperity, that the true system of government consists not in kings but in fair, equal and honorable representation." Paine was, once again, unable to sway the Convention. After Louis Capet's treason conviction, in January 1793, the Convention voted, by a slim margin, for the death penalty. The decision filled Paine "with genuine sorrow" and he urged the Convention to postpone the execution. Paine did not speak French, so he had his speeches translated and read aloud by another deputy. As his translator attempted to share Paine's views with the Convention, the radical physician and journalist Jean-Paul Marat shouted him down, proclaiming Paine a Quaker, opposed on religious grounds to capital punishment. But shouts of "free speech" briefly prevailed, and Paine's position was made plain. His objections were not religious. They were constitutional. The punishment of the King, he explained, was not the duty of a body convened to create a new constitution. The King had been dethroned; his guilt established. Let the business of constituting a new government go forward and let that new government administer justice, in accord with the law.

Paine also pointed to the international consequences of regicide. "France's sole ally is the United States," he reminded his colleagues. "Now, it is an unfortunate circumstance that the individual whose fate we are at present determining has always been regarded by the people of the United States as a friend to their own revolution. Should you come, then, to the resolution of putting Louis to death, you will excite the heartfelt sorrow of your ally." Again, Marat silenced Paine's translator: "I denounce the translator. Such opinions are not Thomas Paine's. The translation is incorrect." In what was perhaps the most dramatic moment of his political life, Paine ascended the tribune from which his translator spoke to proclaim that, no, indeed the translation was correct. These were his words. The translator then continued, "I beg that you delay the execution. Do not, I beseech you, bestow upon the English tyrant the satisfaction of learning that the man who helped America, the land of my love, to burst her fetters, has died on the scaffold."[1]

Paine's pleas, of course, came to naught. The Revolution, it seemed, would fulfill the prophecy of its greatest critic. With the King's execution, on January 21, Burke lamented, "the Catastrophe of the tragedy of France has been completed." But there was no solace in vindication for Burke. "I looked for something of that kind as inevitable," he explained to a friend, "yet, when the fatal, and final Event itself arrived, I was as much leveled, and thrown to the Ground in the consternation, as if it were a thing I had never dreamt of."[2]

In an ill-advised moment of candor some months after Louis's execution, Paine wrote Danton, "I now despair of seeing the great object of European liberty accomplished, and my despair arises not from the combined foreign powers, not from the intrigues of aristocracy and priestcraft, but from the tumultuous misconduct with which the internal affairs of the present revolution are conducted."[3]

BY THE FALL of 1793, the systematic denunciations and judicial murders known as the Terror were in full force. Paine himself had been repeatedly denounced but survived the previous summer and much of the fall by working in the service of the ruling Jacobin faction. He had devoted himself to a series of schemes, including one to import desperately needed flour from the United States and another to acquire saltpeter from British vessels returning from India. On Christmas Eve, after months spent watching one friend after another led off to their deaths, Paine was arrested and taken to the Luxembourg prison. The fifty-six-year-old Paine would spend the next ten months there, narrowly escaping death from both illness and the executioner.

When his health permitted, Paine campaigned for release. He challenged French officials on the grounds that he was a U.S. citizen, imprisoned with neither cause nor due process. He begged friends to take up his cause, but his pleas achieved little. American and British friends in Paris feared the wrath of a Jacobin regime rooting out alleged foreign agents. For the United States government, action on Paine's behalf risked angering the Jacobins and undermining relations with Great Britain. Making matters yet more difficult, the American minister in Paris, Gouverneur Morris, had no sympathy for Paine's plight.

Morris, whom Paine had known in Philadelphia, had been an early proponent of Paine's bridge. But when the two met in London in the spring of 1792, as Morris made his way to Paris, their differences over the Revolution erupted into bitter argument. Much like Burke, Morris had long since concluded that revolutionary France was headed toward disaster. As he confided to his diary in the fall of 1790, "this unhappy country, bewildered in the pursuit of metaphysical whimsies, presents to one's moral view a mighty ruin. . . . One thing only seems to be tolerably ascertained, that the . . . Revolution has failed." Paine would later describe Morris's posting to France as "the most unfortunate and injudicious appointment that could be made." His reason was that Morris's views "gave every reason to suspect that he was secretly a British Emissary."[4]

BY THE SPRING of 1794, the Washington administration had begun to see Morris as a liability. With Britain and France at war, and with the United States seeking improved trade relations with Britain, it walked an ever-finer line of neutrality. Rather than risk needlessly angering its oldest European ally, it recalled Morris, replacing him that August with Senator James Monroe of Virginia. Monroe could hardly have been more unlike Morris. A protégé of Jefferson, Monroe was sympathetic to the French republican cause. He opposed rapprochement with Britain and favored strengthening ties between the modern world's new republics. Not long after arriving in Paris, in willful denial of the bloody Terror, Monroe stood before the National Convention to declare his fealty to the Revolution. He celebrated French military achievement and attacked the forces of counter-revolution, especially those surging forth from the British Isles. This performance earned him the happy approval of his French hosts but infuriated his president, whose British envoy, John Jay, was covertly working toward a new Anglo-American trade agreement. For Paine, however, Monroe was a savior.

The new ambassador admired Paine and was determined to gain his release. This was, he assured Paine, the duty of any representative of the American people.

The crime of ingratitude has not yet stained, and I trust never will stain, our national character. You are considered by [the Americans], as not only having rendered important services in our own revolution, but as being, on a more extensive scale, the friend of human rights, and a distinguished and able advocate in favour of public liberty. To the welfare of Thomas Paine the Americans are not, nor can they be, indifferent. . . . To liberate you, will be an object of my endeavours, and as soon as possible.[5]

Paine urged Monroe to challenge French authorities on the definition of republican citizenship. Instead of a status determined
by the state itself, it had to be seen as a matter of voluntary and
fraternal allegiance. In refusing to accept Paine's claims that he
was an American citizen, the French violated these principles.
Monroe adopted a more pragmatic path. Fearing that such
arguments could be misinterpreted as counterrevolutionary, he
argued Paine's case as a simple matter of criminal justice. If in
fact Paine had committed a crime, the time had come for him
to stand before his accusers. If there had been no crime, the
regime was obliged to release him.

The strategy succeeded. On November 6, 1794, Paine was
finally allowed to leave Luxembourg prison. Months in a cold,
dark, and dank room had taken their toll on him, but he was
now, once again, a free man.

After Paine recovered from a painful abscess and a
life-threatening case of typhus, his circumstances improved.
While convalescing in Monroe's Paris home, he was able
to reclaim his seat in the National Convention, along with
the back pay due after his wrongful imprisonment. He also
returned to the political stage with new vigor, urging the new
Thermidorian government, which had ruthlessly purged the
revolutionary government of Robespierre's Jacobins and their
sympathizers, to resist another slide into terror. He would
begin writing again, publishing a brief tract, *Dissertations sur
les Premiers Principes de Gouvernement*, defending the democratic principles he had so successfully outlined in *Rights of
Man*. But Paine's first concern was finding a way back to his
American home, a goal that had eluded him before his imprisonment and would continue to do so for the next seven years.

AFTER SIGNING A TRADE treaty with Great Britain in late November 1794, the Washington administration abandoned any pretense of neutrality in the Revolutionary Wars. In response, French privateers and naval ships began attacking American vessels, and by 1798 the United States and France were engaged in the "Quasi-War," an undeclared naval struggle that would last until 1800.

Trapped in a foreign country, Paine spent his days contemplating the course of revolution and receiving visitors in a rented Left Bank room at 4 rue du Théâtre Français. One of those visitors, the Irish reformer Wolfe Tone, recalled Paine as "conscientiously an honest man," but he also remembered a conversation with Paine about their mutual friend, Edmund Burke: "I mentioned to him that I had known Burke in England, and spoke of the shattered state of his mind, in consequence of the death of his only son Richard. Paine immediately said that it was the 'Rights of Man' which had broken his heart." In the privacy of his diary, Tone responded with disgust: *"Paine has no children!"*[6]

IN THE FALL of 1798, English newspapers had begun carrying stories about Paine's return to architecture. "Thomas Paine," they reported, "tired or disgusted with politics has been employed these two years in the construction of an iron bridge." By 1802, that work had led to two new bridge models. Henry Redhead Yorke, an Englishman who visited Paine in Paris that year, recalled that the American "seems to have a contemptuous opinion not only of books, but of their authors; for in shewing me one day the beautiful models of two bridges he had devised, he observed that Dr. Franklin once told him that 'books are written to please, houses built for great men, churches for priests, but no bridge [is built] for the people.'" Once again, Paine was building bridges for the people. Paine's new bridge models, Yorke remembered, "exhibit an extraordinary degree not only of skill, but of taste, in mechanics; and wrought with extreme delicacy, entirely by [Paine's] own hands." So fine were the models, Yorke recalled, someone offered to purchase them for the vast sum of three thousand pounds.[7]

As long as Paine remained in France, the models would be little more than curiosities. Whatever access he had had to the French architectural establishment had long since been lost. The aged Perronet died shortly after Paine was imprisoned and the French scientific and architectural world had been reshaped by revolution. In sweeping away the royal schools and guilds that governed knowledge in prerevolutionary France, the revolutionaries also swept away the fluid, cacophonous world of invention and philosophy that had been so welcoming of Paine. The world of Franklin had given way to the world of Napoleon, the revolutionary general who would begin ruling France in November 1799. And what Napoleon stood for was bureaucracy, regulation, and state power. In place of the court favorites and aristocratic institutions of the old regime came a new regime of professionalism, meritocracy, and order.

A new École Polytechnique would prepare the French republic's engineers. Admissions to the École would be determined by new nationwide exams. Students would begin their training with rigorous studies in mathematics and geometry, and only the most able graduates would move on to the École des Ponts et Chaussées. There, they would encounter a quasi-military culture, with ranks, uniforms and even the signature ritual of the honor-driven, martial world: the duel. There was no room in this new professional order for the self-taught bridge builder.

PAINE'S architectural prospects outside of France were little better. He still had many admirers in America and Britain, but that admiration had grown quiet, in part because of Paine's apparent support of the French Revolution, even after its turn to terror. Worse for Paine though was a terrible political miscalculation: he had wrongly concluded that President Washington was directly responsible for his long imprisonment. Although Morris was probably more to blame than anybody else, Paine published an open letter attacking Washington and his administration. Parts of the letter appeared in American newspapers in the fall of 1796. Paine condemned the perfidy of the U.S. government and attacked the President himself as a cowardly hypocrite. Consider the fact, Paine urged his readers, that the man who so publicly celebrated the deep and abiding Franco-American alliance was all the while seeking trade deals with Britain. "In what a fraudulent light must Mr. Washington's character appear in the world," Paine exclaimed, "when his declarations and his conduct are compared together!"[8]

Washington's reputation was nowhere near as peerless as it had been when he had taken office in 1789. Nonetheless, Paine paid dearly for his attack. In a typical condemnation, one Mas-

sachusetts newspaper proclaimed it the "most extraordinary
composition of abuse, petulance, falsehood and boyish vanity
that ever came from Grub Street or a garret." Even this attack
on the character of America's most esteemed public figure,
however, paled in its consequences next to Paine's pamphlet
The Age of Reason. Several years in the making, *The Age of
Reason* was first published in English in early 1794, not long
after Paine's imprisonment.[9]

The tract was among the fiercest attacks on organized
Christianity ever written. One could not read the Bible, Paine
wrote, with its "obscene stories," its "voluptuous debaucheries,
the cruel torturous executions, the unrelenting vindictiveness,"
without concluding that it "would be more consistent that we
called it the word of a demon than the Word of God. It is a
history of wickedness that has served to corrupt and brutal-
ize mankind; and, for my part, I sincerely detest it as I detest
everything that is cruel." Much like the old monarchical governments Paine had so long decried, so the defenders of this
Biblical history, often acting at the behest of those very governments, served themselves before they served their followers. This was contrary to the ideals of Paine's Jesus. "Out of
the matters contained in" the Bible, "the Church has set up a
system of religion very contradictory to the character of the
person whose name it bears. It has set up a religion of pomp
and of revenue, in pretended imitation of a person whose life
was humility and poverty."[10]

Although *The Age of Reason*, at least as Paine understood
it, was ultimately a defense of true religion, Paine's enemies
saw it as an atheist manifesto. At a time when even moderates
feared associations with radical Jacobinism and its ruthless

anticlericalism, such fringe religious views became particularly hard to forgive. Henceforth, there were few references to Paine not prefaced with the word "infidel." Architecture, even for the democratic Paine, had always been dependent on the patronage of the wealthy and the powerful. Now, few social elites on either side of the Atlantic would associate with him. Those British architects who admired him as architect and pamphleteer would begin to conceal their views. Thomas Telford, the great Scottish-born bridge builder, sometimes referred to as the Colossus of Roads, was typical.

BY THE MID-1790S, CONGESTION on the Thames threatened London's mercantile supremacy. The massive West India Docks and the London Docks downstream had not yet been built and London Bridge prevented oceangoing vessels from traveling upriver to the city's growing mercantile heart. Among the proposed solutions was a new London Bridge that would allow high-masted ships to pass easily to the quays and warehouses of the City. In 1799, Parliament established a committee to review proposals for addressing Thames congestion. The most majestic of these, and an essential addition to the catalog of the world's most remarkable unbuilt structures, was a 600-foot iron-arch bridge proposed by Telford and James Douglass, another Scottish architect.

Telford had long been an admirer of Paine's *Rights of Man*. The book, which appeared when Telford was thirty-three, had persuaded him that "nothing short of some signal revolution can prevent [Great Britain] from sinking into bankruptcy, slavery, and insignificancy." Telford, it would seem, had been the supreme Painite: a believer in revolutionary reform and a builder of bridges. But when he solicited expert endorsements for his proposed London bridge, he came to see just how incompatible Painite radicalism and architecture had become. Among the experts Telford consulted were the Walkers of Rotherham, one of whom replied that the

> idea of an arch of 600 feet is a bold stretch—Tom Paine's opinion was, that a Bridge upon his System might be thrown over the Atlantic, if centres for erecting it could be fixed. We have however liv'd to see his System of Bridges, as well as of Politics, exploded.

Telford responded to the Walkers accordingly. He disclaimed any knowledge of Paine or his bridge, adding a jovial stab at the "political Quixote." Still, he had to acknowledge that whatever Paine's political failings, the old revolutionary was right to assume that the future of bridge building lay in iron, a substance that would permit "the span of arches" to be "extended considerably farther than by the present practice."[11]

In early 1791, when Telford read *Rights of Man*, the work was a sensation in Britain. In Sheffield, Manchester, and London, artisans, mechanics, and radical intellectuals had joined together in committees of correspondence and constitutional reform associations to discuss Paine's ideas. Those discussions gave rise to hundreds of less well-known pamphlets, calling for constitutional reform at home and celebrating the achievements of revolutionaries abroad. Within two years, the government's crackdown would dampen Painite enthusiasm, but a declaration of war by France, the Jacobin terror, and Paine's attack on Christianity did far more to quiet Paine's British disciples. By 1800, the year the Parliamentary committee evaluated Telford's Thames River bridge, no serious promoter of public works dared acknowledge a debt to Paine.

PAINE CONTINUED to believe that invention would ultimately triumph over politics and personal attack. Whatever the world thought of him, it recognized the universal genius of his invention. Such, in any event, was the reasoning behind his determination to return to America and resume his bridge building. Paine's prospects for returning brightened when his friend and ally, Thomas Jefferson, was elected to the presidency in 1800. With the 1802 Treaty of Amiens, which momentarily ended hostilities between revolutionary France and its enemies in Europe, Paine could finally make safe passage to his home nation. On September 1, 1802, the sixty-five-year-old Paine boarded a ship at Le Havre and set sail for the United States.

The temporary end of the French Revolutionary Wars in Europe also made possible another Atlantic crossing, one Paine knew nothing of, but one he had made possible. Sometime in 1801, a ship set sail from Hull, England, bearing an 87-ton iron-arch bridge. The structure had been fabricated at the Walker brothers' Rotherham works and was to be erected across the Rio Cobre near Spanish Town, Jamaica, the colony's capital city. The bridge would link Spanish Town to Jamaica's new commercial capital, Kingston. Erected in 1802, it remains in place to this day, making it the oldest substantial example of iron architecture in the Western Hemisphere. The Spanish

Town structure also stands as a monument to one of Paine's least recognized innovations. As he had noted in his 1788 British patent application, iron construction turned architecture into an exportable commodity. Iron members cast in Britain could be sent anywhere in the world.

In 1819, British authorities erected a naval hospital in Port Royal, Jamaica, also with a prefabricated iron structure sent from England. Jamaican builders had turned to iron because the Caribbean island had been deforested. In the United States, the situation was entirely different. The abundance of old-growth forests made the country no place for a builder of iron bridges.[12]

13

No Nation of Iron Bridges

O<small>N</small> O<small>CTOBER</small> 30, 1802, after a two-month passage, Thomas Paine landed in Baltimore with his Paris bridge models. Several days after, he made the short journey to the nation's new capital, Washington City. There, he deposited his bridges at the U.S. Patent Office, housed in the State Department. Paine then took up residence in one of the city's rooming houses and began composing a petition for Congress, later published as his essay "The Construction of Iron Bridges." The document explains that Paine was not seeking personal gain. He had no intention of taking an American patent on his invention; he simply wanted his adopted homeland to see the revolutionary promise of iron-arch construction. All that was needed, to this end, were funds for a single iron supporting rib of four hundred feet. "It is an advantage peculiar to the construction of iron bridges," Paine explained, "that the success of an arch of a given extent and height can be ascertained without being at the expense of building the bridge." Paine would oversee the design and construction of the structural rib and, upon demonstrating its capacity to sustain weight, would devote himself to the construction of the full Schuylkill bridge.[1]

The Congressional petition was more than a plea for money; it was also an act of self-defense. The author of *Common Sense*, *Rights of Man*, and *The Age of Reason* had not returned to his country to foment revolution or to argue the principles of Christianity. He had come as an agent of technological change. Other Americans might be building stone or wood bridges, or have the skill to craft the parts of an iron bridge, but few had the knowledge to assemble those parts into a structure that would make America's rivers fully integral arteries of the new nation's commercial life. In the United States, Paine alone was so equipped. With Congressional patronage to reassure investors, an iron bridge would soon be built across the Schuylkill.

In the end, the prospects for the petition were poor. Paine knew that his reputation had suffered terribly but he had no sense of the full depth of anti-Paine sentiment. Before Paine had even stepped ashore, Jefferson's Federalist opponents attacked the President for supporting his return. What are the "religious or even decent, in this country, to think," wondered the Baltimore *Republican*, when their president seems "determined to honor the drunken, impious wretch," Tom Paine? With Paine's arrival in Baltimore, the same paper urged the city's lowly Jeffersonians to carry Paine about town in a cage. For, "no one would grudge fifty cents to see the demi-human arch beast" and "a very considerable sum might be collected— enough with which Paine might get drunk daily as long as he lives." President Jefferson, it seemed, had refused to recognize the unpleasant reality that Paine was no longer the great American he had known many years earlier. "Paine's character in

Paris," the *Virginia Gazette* observed, "is well known to have been so despicable . . . that his company was avoided, like the presence of a person infected with the plague. . . . Frequently he was found rolling and tossing in the streets, in a state of intoxication that astonished the Parisians unused to such spectacles of human depravation."[2]

"We cannot suppose that this flaming comet," the *New York Evening Post* noted, referring to Paine, "whose fiery course has so long astonished and terrified mankind will now stop its career, but rather that it will acquire accelerated rapidity in its orbit, and catch additional fire from its near approach to that sun of political, moral, and religious virtue, Mr. Jefferson." Jefferson's fame would bring new vitality to the former revolutionary, now better known for his face, "very red, and covered with a plentiful quantity of liquor blotches" and a nose "uncommonly large, red, and carbuncled." The dangers Paine posed were on the minds of the editors of the *New York Gazette*, who reported that upon Paine's arrival at a rooming house in Georgetown,

the landlord ushered him into the supper room with the formal introduction of 'gentlemen, *this is Mr. Thomas Paine*, author of Common Sense, The Crisis, Age of Reason, & C.' when the company almost to a man rose from the table, and one of them in the name of the rest, called the landlord aside and informed him if Mr. Paine was permitted to stay in his house, they should consider it as their passport to quit it—When the Landlord, rather than loose a number of gentlemen, whispered in the ear of *Mr. Jefferson's affectionate friend* that it would be somewhat

inconvenient for him to accommodate him, and requested
he would seek other quarters, which he was constrained
to do.[3]

This likely apocryphal story nonetheless captured political
reality and explains why Jefferson refused to take his friend's
bridges before Congress. After less than two months in Washington, a disgusted Paine shipped the models north to Philadelphia. The President had "not only shown no disposition
towards" pursuing the bridge project, he "by a sort of shyness
. . . precluded it." His friend, Paine believed, had capitulated
to partisan spirit, denying Americans an obvious public good.[4]

WITH NO HOPE for federal government support, Paine returned
to Philadelphia. But his prospects there were equally poor. The
painter Charles Willson Peale installed Paine's models in his
Philadelphia Museum, where they would remain, displayed
alongside mastodon bones, stuffed birds, works of art, and
other curiosities. The network of friends and fellow inventors
who had long encouraged Paine was now gone.

Benjamin Franklin and David Rittenhouse had died, and
many of those who remained wanted nothing to do with
Paine. His fellow revolutionary the evangelical Benjamin Rush
refused to see him. "His principles," Rush explained to a friend,
"avowed in his 'Age of Reason,' were so offensive to me that I
did not wish to renew my intercourse with him." The reception
was so hurtful that Paine spent only two days in the city. In the
years that followed, the collapse of his reputation there would
be a continued source of personal distress. In early 1806, after

he discovered that the mayor of Philadelphia had publicly slandered another man by accusing him of being a disciple of the hated Paine, the city's most celebrated revolutionary replied to the mayor with the dismissive, "I set too much value on my time to waste it on a man of so little consequence as yourself."[5]

EVEN IF Paine's reputation had survived the years of the French Revolution, the prospects of his new bridge would have been poor. While he built his models in Paris, American architects were devising their own much cheaper methods for addressing the country's riverine barriers. Those new methods rested on a simple truth: what America wanted were inexpensive and easily constructed bridges. Stone and iron bridges were neither. In his own proposal for a Schuylkill bridge, Charles Willson Peale captured the sentiment succinctly. "I offer you," he explained to readers of his 1797 pamphlet on wooden bridges, "a cheap and easy mode of building bridges, the principles of which are so simple, and the mechanism so easy, that any ingenious man may execute them." Peale proposed a bridge whose supporting arches were composed of multiple layers of laminated wooden planks. Any reasonably skilled carpenter could build such a bridge.[6]

In 1811, Thomas Pope, an obscure craftsman and architect from New York City, published the first systematic American treatise on bridge building. The book surveyed the history of world bridges, although it made no mention of Paine. The reason, perhaps, was that Pope too saw no need to build with anything other than what the American countryside provided in so much abundance. Through lengthy calculations and tedious scientific argument, Pope established that wood was the ideal bridge-building material. Experimental science, he explained to his readers, demonstrated that "timber, as to its tension, and also compression, is not that weak and watery substance, so easily affected, as the blind theorists of our time would have them believe." The enemies of wood rested their claims on vague theory, but its champions could show its superiority "without the assistance of any of the *Sir Isaac Newtons* of the present day." The supreme expression of this truth would come from a novel design, Pope's flying pendant lever bridge, a structure whose pedigree lay more with the gothic flying buttress than the classical arch. Much like Paine, Pope believed his bridge could span virtually any river, and he presented to his readers an 1800-foot example designed to cross the East River between Brooklyn and Manhattan. The bridge was a single span that rose to 223 feet above the water. Instead of supporting itself, as any arch would, its lengthy arms would be anchored to the riverbanks by monumental embankment towers. For Pope, there was never any question that his stunning bridge would be made from wood.[7]

Nothing came of either Pope's or Peale's design, but they did reflect the near unanimous conclusion of American bridge builders. The United States would be a country of wooden bridges. In a neoclassical age stamped by the monumental buildings of antiquity, there was a certain ambivalence about this fact. To some, the sacrifice of permanence for ease of construction defeated the entire classical project. In place of great and lasting monuments, Americans made ephemeral buildings, vulnerable to fire, water, and the other forces of nature. As one early critic put it, "Bachelors only ought to build of wood—men who have but a life estate in this world, and who care little for those who come after them." But these kinds of lamentations carried little weight. Americans grew so adept at wood construction that it was often difficult to distinguish from the more lasting forms of the Old World. Upon arriving in New York City during his famous American tour, Alexis de Tocqueville was surprised to notice "At some distance out from the city a number of little white marble palaces, some of them in classical architectural style. The next day, when I looked more closely at one of those that had struck me most, I found that it was built of whitewashed brick and that the columns were painted wood." [8]

There was no greater monument to America's appetite for cheap wooden construction than the first Permanent Schuylkill River Bridge. Begun in 1801 and completed in 1805, the three-arched bridge spanned the river at Market Street. It had been designed and built for the Schuylkill Permanent Bridge Company by Timothy Palmer, a Massachusetts carpenter and self-taught bridge builder. Palmer had built a series of bridges across the Merrimack, Piscataqua, and Potomac Rivers. At the time of its completion in 1794, his bridge across the Piscataqua, near Portsmouth, New Hampshire, was America's greatest bridge. Its central arch, composed of three parallel timber ribs, was 244 feet in length. But Palmer's most celebrated achievement by far would be the Schuylkill Permanent Bridge. The astonishing structure was nearly 500 feet long and consisted of three low-slung wooden arches on masonry piers. Palmer's work had a distinguished pedigree. In the mid-eighteenth century, two Swiss brothers, Johannes and Hans Ulrich Grubenmann, had begun building wooden bridges in northern Switzerland, the most famous of which spanned the Rhine near the small town of Schaffhausen.

As Paine had long recognized, the same qualities that made wood-bridge construction so cost-effective also made it vulnerable. Rain, wind, snow, and ice, combined with the pulverizing effects of traffic, would all eventually induce wood rot. The Grubenmann brothers addressed these problems by enclosing their bridges with roofs and walls. Palmer's bridge was not, initially, enclosed, a choice Palmer himself considered short-sighted. "It is sincerely my opinion the Schuylkill Bridge will last thirty and perhaps forty years, if well covered." The alternative was for "this beautiful piece of architecture . . . which has been built at so great expense and danger, to fall into ruin in ten or twelve years." Palmer's views ultimately prevailed and, despite opposition from profit-minded shareholders, the Schuylkill Permanent Bridge Company hired Philadelphia architect and carpenter Owen Biddle to enclose the bridge. The resulting structure, Biddle observed, was initially "the only *covered* wooden bridge, in any country, except, perhaps," Switzerland. But it would set a lasting American precedent.[9]

For foreign visitors, few structures were as striking as America's covered bridges. Michel Chevalier, a French economist who toured the United States in 1834 observed that "the Americans are unequalled in the art of constructing wooden bridges; those of Switzerland, about which so much has been said, are clumsy and heavy compared to theirs." James Silk Buckingham, an English journalist and Member of Parliament who visited the United States shortly after Chevalier, wrote that American bridges are "generally of wood, and are enclosed with sides and roofs, so that they form long arched tunnels over the streams, with windows on each side for light and air." James Fenimore Cooper, writing as a fictional European bachelor touring the United States, noted that "a bridge, a quarter, a half, or even a whole mile, in length . . . is no extraordinary undertaking for the inhabitants of a country which, forty years before, and often less, was an entire wilderness." For Cooper, "these avenues of timber" defied description. Often "the traveller, perhaps whilst ruminating on the recent origin of this country, finds himself journeying through an edifice which is from a quarter of a mile to a mile in length." Through the course of that journey, according to another American writer, the traveler might read "the inscriptions in chalk and coal of those who had gone before."

Charles Dickens, who visited the United States in 1842, described the extraordinary experience of approaching Harrisburg, which had become Pennsylvania's capital in 1812. Dickens entered the town after passing through the nearly 3000-foot-long Camelback Bridge, named for the undulating appearance of its lengthy, covered structure:

We crossed [the Susquehanna] by a wooden bridge, roofed and covered on all sides, and nearly a mile in length. It was profoundly dark; perplexed, with great beams, crossing and recrossing it at every possible angle; and through the broad chinks and crevices in the floor, the rapid river gleamed, far down below, like a legion of eyes. We had no lamps; and as the horses stumbled and floundered through this place, towards the distant speck of dying light, it seemed interminable. I really could not at first persuade myself as we rumbled heavily on, filling the bridge with hollow noises, and I held down my head to save it from the rafters above, but that I was in a painful dream; for I have often dreamed of toiling through such places, and as often argued, even at the time, "this cannot be reality."[10]

AFTER PALMER, the German émigré Lewis Wernwag became the new nation's most celebrated bridge architect. Catapulted to prominence by a building boom that followed the War of 1812, Wernwag designed and built twenty-nine mid-Atlantic bridges. His most famous, and among his first, was the Lancaster–Schuylkill Bridge, completed in the spring of 1813. Known as the "Colossus," the 340-foot single-arch covered bridge crossed the river at Upper Ferry, replacing a dilapidated old floating bridge. The distinctive Greek Revival covering was designed by Robert Mills, an architect who had studied with Thomas Jefferson and the genius of American neoclassicism, the English immigrant Benjamin Henry Latrobe. Mills's portfolio consisted mostly of elegant Greek Revival churches, including the Sansom Street Baptist Church and the octagonal First Unitarian Church, both in Philadelphia. Later, in 1845, Mills would receive his best-known commission, the Washington Monument.

The "Colossus" was immortalized in countless prints, depicted on Staffordshire dinnerware, and captured the imagination of every traveler to encounter its majestic form. Upon beholding the bridge, the English actress Fanny Kemble wrote that "at a little distance, it looks like a scarf, rounded by the wind, flung over the river." In 1838, the

bridge succumbed to a well-known hazard of wooden architecture. Within minutes, it was destroyed by an arsonist's
torch. "The burning bridge, the crowd upon the hills, and
the beautiful scenery in the neighborhood, illuminated by
the blaze" reported Philadelphia's *Public Ledger,* "presented
a splendid picture."[11]

The year after the "Colossus" burned, builders completed
a bridge across Dunlap's Creek in Brownsville, south of Pittsburgh. The bridge carried the National Road, which linked
Baltimore to Wheeling, West Virginia, and eventually drew
together towns across Ohio, Indiana, and Illinois. It was also
the first iron-arch bridge built in America. It still stands, carrying what is now U.S. Route 40. Perhaps the builders hoped
to avoid the fate of the "Colossus." Or perhaps, after an earlier
wooden bridge collapsed in a snowstorm, they simply decided
that the time had come for an American iron bridge.

DEFEATED BY the city that had once done so much to nourish
his mind and spirit, in late February 1803, Paine retreated to
the home of friends in Bordentown, New Jersey. In the fall, he
made his way to his farm in New Rochelle. He hoped to make
ends meet by harvesting firewood but his sixty-six-year-old
body made this impossible. He spent the winter doing little but
fending off the cold and solitude of rural life. The next year, as
winter descended, Paine abandoned the countryside for New
York City. During the next few years, he would travel back and
forth between New Rochelle and New York, suffering increasingly from fits of gout and assorted other symptoms of age.
Though he was able to cover some debts and briefly generate

income from the sale of part of his New Rochelle property, he was nearly destitute.

A lifetime of refusing remuneration for his writings and building bridges on his own dime had culminated in an old age of terrible poverty and infirmity. Paine's plan to alleviate precisely this kind of suffering, carefully laid out in *Agrarian Justice*, his final substantial pamphlet, was a distant dream in the United States. Written in the winter of 1795–1796, a little over a year after Paine's release from Luxembourg prison, the pamphlet railed against human inequality and was among the first works of political economy to explain poverty as an outgrowth of "that which is called civilized life," rather than a "natural state." Paine observed that the Native peoples of America would never recognize the impoverishment of Europe's poor because "civilization," whose abundant advantages include "agriculture, arts, science, and manufactures . . . has operated in two ways: to make one part of society more affluent, and the other more wretched." The aged, the infirm, children, military veterans, and all the others among society's disadvantaged had been left behind by the progress of civilization and a chasm now divided society. The solution, for Paine, was taxation. By imposing an inheritance tax on all landed estates, he argued, dynastic wealth would be curbed and society's most disadvantaged would get the basics that were rightfully theirs.[12]

Agrarian Justice was written at a hopeful moment for Paine. The ruthless Jacobin regime had been destroyed and a new regime, the Directory, named for its five-person executive committee, controlled France. Meaningful social reform seemed possible. But now, in the early years of the nineteenth century, Paine found himself in a United States with little regard for either its aged poor or its political forebears.

Perhaps the ultimate insult came in the fall of 1806, when Paine returned to New Rochelle to cast his vote in local elections. To his utter horror, the elections supervisor refused to accept the vote on the grounds that Paine was not a citizen of the United States. Even in the village of New Rochelle, there was political gain to be made at the expense of Thomas Paine. Paine hired an attorney to challenge the outrageous claim and began lobbying former New York governor and now Vice President George Clinton to attest to his citizenship. But the case went nowhere. No judge or jury in New York would declare the infidel Paine an American. Paine's treatment remains astonishing, especially given that his property—the basis for his right to vote—had come directly from New York State itself.

Over the course of the next three years, Paine's health deteriorated and he stopped going to his farm. His plans to publish his collected works were set aside for want of money and physical capacity. Much as he had in Philadelphia thirty years earlier, he attempted to work his way into New York's political and literary scene. He published several essays on New York State politics and the naval vulnerabilities of the Port of New York. But these did little for his reputation or his spirits. By early 1808, he was living in a wretched garret on what is now Fulton Street in lower Manhattan. Having spent so much of his time seeking compensation for his service to the United States, and having been repeatedly insulted and rebuffed since his return from France, Paine now refused to seek help. Charity, he had come to believe, was just another instrument through which the powerful held sway over the powerless. A human being's capacity to live a life of reasonable health and happiness, he had written in *Agrarian Justice*, should not be left to the whims of well-to-do philanthropists. For it was "but a right, not a bounty but justice," and by definition, no single person, purely owing to the fortunes of birth or industry, should have the power to determine who should have food on her plate.

> The present state of civilization is as odious as it is unjust.
> It is absolutely the opposite of what it should be, and it
> is necessary that a revolution should be made in it. The
> contrast of affluence and wretchedness continually meet-
> ing and offending the eye, is like dead and living bodies
> chained together.[13]

By the summer of 1808, it was becoming clear that Paine could no longer survive on his own. Friends arranged to have him moved from his rented Fulton Street room to the home of Mr. and Mrs. Cornelius Ryder in Greenwich, then a separate village a couple of miles north of the city.

The arrangement made for a brief improvement, but during the fall Paine's condition worsened and by January 1809, the end was near. Paine prepared his final will and testament but death did not come quickly. Over the course of the next months, he suffered with agonizing pain, festering sores on his feet and ankles, and incontinence. It was a miserable decline that finally ended on the morning of June 8. Madame Bonneville, a French émigré whom Paine had befriended in Paris, and whose children he had adopted, accompanied his casket to his farm in New Rochelle. She and her son attended to Paine's burial and ordered a small tombstone for the private burial plot. Its brief inscription read: "Thomas Paine, Author of 'Common Sense,' died the eighth of June, 1809, aged 72 years."

In the years to come, Paine's gravesite attracted visitors, all seeking a piece of the great American revolutionary. "Many persons have taken away pieces of the tombstone and of the trees" planted nearby, Madame Bonneville recalled. "Foreigners especially have been eager to obtain these memorials, some of which have been sent to England. They have been put in frames and preserved [and] verses in honor of Paine have been written on the head stone."[14]

conferring control of cities and towns on small, nepotistic governing boards, these legal agreements extended the privilege of the few at the expense of the many. For this reason Pennsylvanians abolished the Municipal Corporation of Philadelphia in the early days of the Revolution. In England, Paine believed, "the generality of corporation towns," which included the city of London, "are in a state of solitary decay." The most industrious Britons chose to live elsewhere, in flourishing new cities such as Manchester, Birmingham, and Sheffield, where government was unconstrained by ancient charters.[1]

But corporations could be beneficial, even when they yielded profits for private interests. Unlike municipal corporations, banks and bridge companies, as Paine understood them, denied no one rights to self-government. The idea that government—as some would now say—picked winners and losers never much bothered Paine, so long as the winners served some clear public need. What Paine never accepted was the idea that even while serving public needs, corporations could become rivals to democratic government. For many Pennsylvanians, the ends simply could not justify the means. The choice between democracy and internal improvement was no choice at all.

In 1786, Paine had dismissed similar attitudes about the Bank of North America. Western Pennsylvanians, he was convinced, attacked the bank not because it was undemocratic but because it challenged the supremacy of their Baltimore trading partners. Whether Paine was correct is difficult to say. But what is certain is that suspicion of corporate power only grew in eighteenth and early-nineteenth-century Pennsylvania, a state that joined the nation as a whole in its delight in conferring

Epilogue

IN THE DECADE AFTER Paine's death, the United States ex⸱
rienced an internal improvement boom. The country b⸱
dozens of bridges and hundreds of miles of roads and cana
Pennsylvania did more to improve itself than any other st⸱
with the exception of New York. But contrary to Thom⸱
Paine's expectation, far from bringing the state closer togeth⸱
all this construction left it more divided. There are many exp⸱
nations for Pennsylvania's persistent sectionalism, too many
survey here.

But one is worth considering. Corporations, chartered
governments, were legal entities that Paine knew well from ⸱
days battling for the Bank of North America, and he had cle⸱
views about this vestige of royal patents and charters. Wh⸱
used improperly, the law of incorporation could harm the g⸱
erned; when used according to the dictates of republican g⸱
ernment, it could serve the public good. The former, for Pain⸱
was most clearly embodied in the municipal corporation.

charters and their associated privileges—to pool capital, issue
stock, raise funds through lotteries and other devices, and gain
monopolistic control of moneymaking enterprises. By 1780,
the former American colonies had granted a mere 7 charters to
private businesses. In the 1790s, they granted some 295. The
trend continued into the nineteenth century. By 1800, the state
of Pennsylvania had chartered 19 corporations. From 1801 to
1815, it chartered 192. The vast bulk of those charters—130—
were for private concerns undertaking public improvements:
canals, turnpikes, bridges, and waterworks.[2]

This charter boom had familiar origins. As the American
states struggled to balance commercial aspiration with fiscal
reality, they faced the same kinds of problems Congress and
imperial Britain had faced decades earlier. Americans were, on
the whole, no more receptive to taxation by their own governments than they had been to taxation by the British government.
For Philadelphia, this fact created a political dilemma. The city
faced continued competition from Baltimore, whose economic
fortunes would soar with the completion of the first leg of the
new National Road in 1818. At the same time, New York was
itself booming. Beginning in 1807, Robert Fulton's steamboats
transformed the Hudson River into the nation's busiest inland
waterway. With the 1825 completion of the 363-mile Erie
Canal, New York City's commercial status would, once and for
all, eclipse Philadelphia's.

The idea that Pennsylvania could somehow tax long
aggrieved western farmers, even to pay for the kinds of
improvements that might improve the economic circumstances
of the whole, was pure political fantasy. If they ever needed a
reminder of this stark truth, Pennsylvania's political leaders

could recall their state's reaction to a federal excise tax on distilled spirits. The Whiskey Rebellion erupted in western Pennsylvania in the summer of 1794, and at one point drew seven thousand Pennsylvanians together in an antitax assault on the small town of Pittsburgh. The ordeal of the regional excise supervisor John Neville would have resonated among any of those who remembered the trials of Philadelphia's Stamp Tax collectors thirty years earlier. For advocates of internal improvement, the revolt, which President Washington put down with federal troops, was an acute reminder of just how unlikely it was that state governments would be able to rely on tax revenue to fund improvements.

With little chance for direct state financing, the best hope for Pennsylvania's economic future lay in the public–private marriage made possible by the law of incorporation. Only by allowing private companies to build and manage their state's bridges, to dredge and control its rivers, to build and maintain its roads, could Pennsylvania's eastern businessmen hope to compete with their neighbors in Baltimore and New York. But since the businessmen most likely to benefit from these improvements were also the Philadelphia financiers most likely to finance them, internal improvement remained divisive.

The problem was evident in 1790, as the state drafted a new constitution. Two western representatives to the convention, James McLene of Franklin County and Andrew Henderson of Huntingdon County, argued for a clear constitutional statement that "perpetuities and monopolies are contrary to the nature of a republican government, and ought not to exist." Not only did corporate charters grant exclusive moneymaking privileges, they also became a species of property, handed down from one generation to the next. This was contrary to the spirit of the American republic.

In the vote on McLene and Henderson's proposal, which did not carry, Pennsylvania's sectional divide was clear: ten of twelve supporters hailed from counties west of the Susquehanna. Of the forty-four delegates opposed, thirty hailed from eastern counties. If this constitutional argument is any indication, the legal dispensation that would allow government to finance a more unified Pennsylvania was dividing Pennsylvanians along old sectional lines.[3]

This sectional division is one reason it took Philadelphia more than a half a century to address its Baltimore problem. The Chesapeake and Delaware Canal, a fourteen-mile waterway linking the two bays, was finally completed in 1829. But the canal was an exception that proved the rule: it was built and financed not by Pennsylvanians, but by the federal government. And it was finished too late to save Philadelphia. By 1830, the city's population had fallen behind both New York's and Baltimore's. Philadelphia would begin to regain its mid-Atlantic primacy in the 1840s, when a new public–private partnership created Pennsylvania's railroads. But the city would never regain the dominance it enjoyed when Thomas Paine arrived there in the fall of 1774.

Notes

About the Sources

In addition to the archival documents, the historical and architectural studies, and the collections of writings of Paine and his contemporaries acknowledged in the notes that follow, I have drawn from Paine's many biographies. The two most authoritative remain Moncure Conway's *The Life of Thomas Paine: With a History of His Literary, Political, and Religious Career in America, France, and England*, 2 vols. (New York: G. P. Putnam and Sons, 1908); and John Keane, *Tom Paine: A Political Life* (New York: Grove Press, 1995). Also excellent are Jack Fruchtman's *Thomas Paine: Apostle of Freedom* (New York: Four Walls, Eight Windows, 1994); and Craig Nelson *Thomas Paine: Enlightenment, Revolution, and the Birth of Modern Nations* (New York: Penguin, 2006). Two other valuable biographical studies are Eric Foner's classic *Tom Paine and Revolutionary America, Updated with a New Introduction* (New York: Oxford University Press, 2005); and Mark Philp's brief life, *Thomas Paine* (New York: Oxford University Press, 2007).

Introduction

1. Jack Whitehead, *The Growth of St. Marylebone and Padding-ton* (London: J. Whitehead, 1989), 10.

2. Philip S. Foner, ed., *The Complete Writings of Thomas Paine* (New York: The Citadel Press, 1945), 2:1053–54 (hereinafter cited as *CW*).

3. Ibid., 2:1411–12.

4. Morris quoted in Foner, "Introductory Note," ibid., 1:xviii; John Adams to Benjamin Waterhouse, October 29, 1805, in Worth-ington C. Ford, ed., *Statesman and Friend: Correspondence of John Adams and Benjamin Waterhouse, 1784–1822* (Boston: Little Brown, 1927), 31; Jacques-Pierre Brissot de Warville, *New Travels in the United States of America, 1788*, ed. Durand Eche-verria (Cambridge, Mass.: Harvard University Press, 1964), 258.

5. Adams to Waterhouse, October 29, 1805, in *CW*, 1:45, 50; Thomas Jefferson to Francis Eppes, January 19, 1821, in Paul Leicester Ford, ed., *The Writings of Thomas Jefferson, 1816–1826* (New York: G. P. Putnam and Sons, 1899), 10:183.

6. L. H. Butterfield, ed., *Adams Family Correspondence* (Cam-bridge, Mass.: Belknap/Harvard University Press, 1963), 1:363.

7. Bernard Bailyn, ed., *The Debate on the Constitution: Federalist and Antifederalist Speeches, Articles, and Letters During the Struggle over Ratification* (New York: Library of America, 1993), 171.

8. Jacob E. Cooke, ed., *The Federalist* (Middletown, Ct.: Wes-leyan University Press, 1961), 86–87. On internal improve-ment, see also Daniel Walker Howe, *What Hath God Wrought: The Transformation of America, 1815–1848* (New York: Oxford University Press, 2007); John Lauritz Larson, *Internal Improvement: National Public Works and the Promise of Pop-ular Government in the Early United States* (Chapel Hill: Uni-versity of North Carolina Press, 2001); and Carol Sheriff, *The Artificial River: The Erie Canal and the Paradox of Progress, 1817–1862* (New York: Hill and Wang, 1996).

9. George Wilson Pierson, *Tocqueville in America* (repr., Baltimore: Johns Hopkins University Press, 1996), 592.

10. *CW*, 1:24.

Chapter 1: River City

1. Thomas Paine to Benjamin Franklin, March 4, 1775, in *CW*, 2:1130–31.

2. Adam Smith, *An Inquiry into the Nature and Causes of the Wealth of Nations*, ed. Kathryn Sutherland (Oxford: Oxford University Press, 1993), book 5, ch. 2, p. 452.

3. *CW*, 2:10, 5–6. On the Excise Service, see John Brewer, *The Sinews of Power: War, Money, and the English State, 1688–1783* (New York: Alfred A. Knopf, 1988); William J. Ashworth, *Customs and Excise: Trade, Production, and Consumption in England, 1640–1845* (Oxford: Oxford University Press, 2003); Thomas P. Slaughter, *The Whiskey Rebellion: Frontier Epilogue to the American Revolution* (New York: Oxford University Press, 1986).

4. Franklin to Hugh Roberts, February 26, 1761, in *The Papers of Benjamin Franklin*, digital edition by the American Philosophical Society, Yale University and the Packard Humanities Institute, http://franklinpapers.org/franklin/framedVolumes.jsp.

5. Gottlieb Mittelberger, *Journey to Pennsylvania*, ed. and trans. Oscar Handlin and John Clive (Cambridge, Mass.: Harvard University Press, 1960), 37; John Hall to Mr. Lilley, January 27, 1768, in "John Hall Letter Book," p. 71, courtesy of Ms. Polly Munson (transcription by Ms. Munson).

6. John Flexer Walzer, "Transportation in the Philadelphia Trading Area, 1740–1775," (PhD diss., University of Wisconsin, 1968), 2–4.

7. Arthur L. Jensen, *The Maritime Commerce of Colonial Philadelphia* (Madison: State Historical Society of Wisconsin, 1963), 8. On the economy of colonial Pennsylvania, see also Clarence P. Gould, "The Economic Causes of the Rise of Baltimore," in *Essays in Colonial History Presented to Charles McLean Andrews by*

His Students (New Haven: Yale University Press, 1931), 224–51; James T. Lemon, *The Best Poor Man's Country: A Geographical Study of Early Southeastern Pennsylvania* (Baltimore: Johns Hopkins University Press, 1972); James W. Livingood, *The Philadelphia–Baltimore Trade Rivalry, 1780–1860* (Harrisburg, Pa.: Pennsylvania Historical and Museum Commission, 1947); D. W. Meinig, *The Shaping of America*, vol. 1, *Atlantic America, 1492–1800* (New Haven: Yale University Press, 1986).

8. Richard B. Sher, *The Enlightenment and the Book: Scottish Authors and Their Publishers in Eighteenth-Century Britain, Ireland and America* (Chicago: University of Chicago Press, 2006), 509–10.

9. Ibid., 531–35.

10. Lawrence A. Peskin, *Manufacturing Revolution: The Intellectual Origins of Early American Industry* (Baltimore: Johns Hopkins University Press, 2003), 42–43, 47–48.

11. John Brewer, *The Pleasures of the Imagination: English Culture in the Eighteenth Century* (New York: Farrar, Straus & Giroux, 1997), 142, 146.

12. *CW*, 2:1131; the essays mentioned appeared in the issues of vol. 1 (April 1775), 158; and vol. 1 (October 1775), 470, respectively.

13. *The Pennsylvania Magazine* 1 (January 1775): 9–11.

14. Ibid. 1 (May 1775): 215; 1 (July 1775): 305; and 1 (September 1775): 416.

15. Ibid. 1 (May 1775): 209.

Chapter 2: The Hazards of Competition

1. Israel Acrelius, *A History of New Sweden: Or, the Settlements on the River Delaware*, trans. and ed. William M. Reynolds (Philadelphia: Historical Society of Pennsylvania, 1874), 165.

2. Gordon S. Wood, *Empire of Liberty: A History of the Early Republic, 1789–1815* (New York: Oxford University Press, 2009), 336–37.

3. *To the Merchants and Other Inhabitants of Pennsylvania* (Phil-

adelphia, 1771); Samuel Rhoads to Benjamin Franklin, May 3, 1771, in *The Papers of Benjamin Franklin*, digital edition by the American Philosophical Society, Yale University and the Packard Humanities Institute, http://franklinpapers.org/franklin/framedVolumes.jsp.

4. *Transactions of the American Philosophical Society* 1, no. 1 (Philadelphia: R. Aitken, 1789): 357–58; James M. Swank, *Progressive Pennsylvania: A Record of the Remarkable Industrial Development of the Keystone State* (Philadelphia: J. B. Lippincott, 1908), 132–33; Rhoads to Franklin, May 3, 1771.

5. Quoted in Jack Rakove, *Revolutionaries: A New History of the Invention of America* (New York: Mariner, 2010), 93.

6. *CW*, 2:20.

7. Benjamin Rush, *A Memorial Containing Travels Through Life* . . . (Philadelphia: Louis Alexander Biddle, 1905), 84.

8. *CW*, 1:13.

9. Richard Gimbel, *A Bibliographic Check List of "Common Sense," with an Account of Its Publication* (New Haven: Yale University Press, 1956), 78–91; Rush, *A Memorial Containing Travels*, 85; Moses Coit Tyler, *The Literary History of the American Revolution, 1763–1783* (New York: G. P. Putnam and Sons, 1897), 1:473; L. H. Butterfield et al, eds., *Diary and Autobiography of John Adams* (1961; repr., New York: Atheneum, 1964), 2:351; Robert A. Ferguson, "The Commonalities of *Common Sense*," *The William and Mary Quarterly* 57 (July 2000): 465–504; Trish Loughran, "Disseminating *Common Sense*: Thomas Paine and the Problem of the Early National Bestseller," *American Literature* 78 (March 2006): 1–28.

10. *CW*, 1:17.

Chapter 3: Years of Peril

1. Thomas Paine to Henry Laurens, January 14, 1779, in *CW*, 2:1164.

2. Ibid., 1:50.

3. Paine to Laurens, January 14, 1779.

4. Dorothy Twohig, ed., *The Papers of George Washington: Revolutionary War Series* (Charlottesville: University of Virginia Press, 1997), 7:397. More generally, on the crossing and its hazards, see David Hackett Fischer, *Washington's Crossing* (New York: Oxford University Press, 2004).

5. Horace Wells Sellers, "Charles Willson Peale, Artist—Soldier," *Pennsylvania Magazine of History and Biography* 38, no. 3 (1914): 276; Caesar A. Rodney, ed. *Diary of Captain Thomas Rodney, 1776–1777* (Wilmington: Historical Society of Delaware, 1888), 1:22–23.

6. Paine to Benjamin Franklin, May 16, 1778, in CW, 2:1143–51.

7. William Spohn Baker, ed., *The Itinerary of General Washington from June 15, 1775, to December 23, 1783* (Philadelphia: J. B. Lippincott,1892), 108.

8. Samuel Hazard, ed., *Colonial Records of Pennsylvania* (Harrisburg, Pa.: Theo. Fenn & Co., 1852), 11:305.

9. Nathanael Greene to George Washington, December 7, 1776, in *The Papers of George Washington*, 7:269; Washington to John Mercereau, April 27, 1777, in ibid, 8:288.

10. Marquis de Lafayette to Washington, December 3, 1777, in ibid., 12:526.

Chapter 4: The Trials of the Republic of Pennsylvania

1. CW, 2:1183.

2. Thomas Paine to Nathanael Greene, September 9, 1780, in ibid,, 2:1189.

3. Paine to George Washington, November 30, 1781, in ibid., 2:1203–4.

4. Ibid., 1:171–85; Paine to Robert Morris, January 24, 1782, in ibid., 2:1205.

5. Eric Foner, ed., *Thomas Paine: Collected Writings* (New York: Library of America, 1995), 312, 310. Also see Alfred Owen Aldridge "Some Writings of Thomas Paine in Pennsylvania Newspapers," *The American Historical Review* 56 (July 1951): 832–38; E. James Ferguson, *The Power of the Purse: A History of American Public Finance, 1776–1790* (Chapel Hill: University of North Carolina Press, 1961); Stephen Mihm, "Funding the Revolution: Monetary and Fiscal Policy in Eighteenth-Century America," in *The Oxford Handbook of the American Revolution*, ed. Edward G. Gray and Jane Kamensky (New York: Oxford University Press, 2013), ch. 18, 327–51.

6. Paine to Daniel Clymer, September 1786, in *CW*, 2:1255–56.

7. *The Independent Gazetteer* (Philadelphia), March 12, 1787, p. 3.

8. *CW*, 1:400; *An Historical Account of the Rise, Progress and Present State of the Canal Navigation in Pennsylvania* (Philadelphia: Zacharian Poulson, 1795), iii.

9. Findley quoted in Bray Hammond, *Banks and Politics in America: From the Revolution to the Civil War* (Princeton, N.J.: Princeton University Press, 1957), 54–55; *The Freeman's Journal: or, North-American Intelligencer* (Philadelphia), March 2, 1785, p. 2; Alfred Owen Aldridge, "Why Did Thomas Paine Write on the Bank?" *Proceedings of the American Philosophical Society* 93 (September 1949): 309–15.

10. *CW*, 2:434.

Chapter 5: The Schuylkill and Its Crossings

1. *The Columbian Magazine* 3 (May 1789): 3-5; Kenneth Silverman, *A Cultural History of The American Revolution* (1976; repr., New York: Columbia University Press, 1987), 604–5.

2. Richard B. Morris, *Government and Labor in Early America* (1946; repr., New York: Harper & Row, 1965); Fred Perry

Powers, "The Historic Bridges of Philadelphia," in *Philadelphia History: Consisting of Papers Read Before the City History Society of Philadelphia* (Philadelphia: The Philadelphia History Society, 1917), 267–316; James M. Swank, *Progressive Pennsylvania: A Record of the Remarkable Industrial Development of the Keystone State* (Philadelphia: J. B. Lippincott, 1908); and John Flexer Walzer, "Colonial Philadelphia and Its Backcountry," *Winterthur Portfolio* 7 (1972): 161–73. On public works and economic development in postrevolutionary and early national Pennsylvania, see Ralph D. Gray, "Philadelphia and the Chesapeake and Delaware Canal, 1769–1823," *The Pennsylvania Magazine of History and Biography* 84 (October 1960): 401–23; Lee Hartman, "Pennsylvania's Grand Plan of Post-Revolutionary Internal Improvement," *The Pennsylvania Magazine of History and Biography* 65 (October 1941): 439–57; Louis Hartz, *Economic Policy and Democratic Thought: Pennsylvania, 1776–1860* (Cambridge, Mass.: Harvard University Press, 1948); Diane Lindstrom, *Economic Development in the Philadelphia Region, 1810–1850* (New York: Columbia University Press, 1978); Wilbur C. Plummer, "The Road Policy of Pennsylvania" (PhD diss., University of Pennsylvania, 1925).

3. *Pennsylvania Packet, and General Advertiser*, April 13, 1784, p. 3; John Hall to Mrs. Capner, January 29, 1786, in "John Hall Letter Book," pp. 63–64, courtesy of Ms. Polly Munson (transcription by Ms. Munson); Jacques-Pierre Brissot de Warville, *New Travels in the United States of America, 1788*, ed. Durand Echeverria (Cambridge, Mass.: Harvard University Press, 1964), 152, fn. 2.

4. G. D. Scull, ed., *The Montresor Journals: Collections of the New York Historical Society, 1881* (New York: New-York Historical Society, 1882), 471.

5. [Thomas Anburey], *Travels Through the Interior Parts of America. In a Series of Letters . . .* (1789; repr. Cambridge,

Mass.: Riverside Press, 1923), 2:170; Samuel Hazard, ed., *Colonial Records of Pennsylvania* (Harrisburg, Pa.: Theo. Fenn & Co., 1851), 7:107, 152; Barr Ferree, ed., *Year Book of the Pennsylvania Society, 1908* (New York: The Pennsylvania Society, 1908), 175–77.

6. *The Pennsylvania Packet or the General Advertiser*, April 10, 1779, p. 3; *Colonial Records of Pennsylvania*, 8:618.

7. Timothy Dwight, *Travels in New England and New York*, ed. Barbara Miller Solomon and Patricia M. King (Cambridge, Mass.: Harvard University Press, 1969), 1:358–59; Brissot de Warville, *New Travels in the United States*, 94, fn 27.

8. Chi Ho Sham, Richard W. Gullick, Sharon C. Long, and Pamela P. Kenel, *Operational Guide to AWWA Standard G300: Source Water Protection* (Denver: American Water Works Association, 2010), 124; *Annual Report of the Water Supply Commission of Pennsylvania for 1914* (Harrisburg, Pa.: Wm. Stanley Ray, 1915), 62; Charles River Watershed Association (www.crwa.org/cr_history.html); Theodore Steinberg, *Nature Incorporated: Industrialization and the Waters of New England* (Cambridge: Cambridge University Press, 1991), 21.

9. Stanley I. Kutler, *Privilege and Creative Destruction: The Charles River Bridge Case* (1971; repr., Baltimore: Johns Hopkins University Press, 1990), 15.

Chapter 6: The Schuylkill Permanent Bridge Company

1. *Minutes of the Philadelphia Society for the Promotion of Agriculture* . . . (Philadelphia: John C. Clark, 1854), 25–28; Brooke Hindle, *The Pursuit of Science in Revolutionary America* (Chapel Hill: University of North Carolina Press, 1956).

2. The full name of the legislation was "An ACT to incorporate the Subscribers to the Plan for erecting a permanent Bridge over the river Schuylkill, at the western extremity of the High

street of the city of Philadelphia." The public debates over a permanent bridge in postwar Philadelphia are treated in "A Statistical Account of the Permanent Bridge, Communicated to the Philadelphia Society of Agriculture, 1806" (Philadelphia: Jane Aitken, 1807), especially 5–18 and 20–21, bound in *Memoirs of the Philadelphia Society for Promoting Agriculture* . . . (Philadelphia: Jane Aitken, 1808), vol. 1; *The Founders' Constitution*, vol. 3, article 1, section 8, clause 7, document 1, available at http://press-pubs.uchicago.edu/founders/documents/a1_8_7s1.html. Before the war, there were at least two proposals to build a permanent bridge, using an iron-chain suspension system, across the Schuylkill near Philadelphia: see Eda Kranakis, *Constructing a Bridge: An Exploration of Engineering Culture, Design and Research in Nineteenth-Century France and America* (Cambridge, Mass.: MIT Press, 1997), 28–34.

3. *CW*, 2; 1051–52; Thomas Paine to Benjamin Franklin, June 6, 1786, in ibid., 2:1027.

4. Andrea Palladio, *The Four Books of Architecture* (1738; repr., New York: Dover, 1965), 69. A concise guide to structural principles is David Blockley, *Bridges: The Science and Art of the World's Most Inspiring Strutures* (New York: Oxford University Press, 2010). Bridge building in Paine's Britain is comprehensively treated in Ted Ruddock, *Arch Bridges and Their Builders, 1735–1835* (Cambridge: Cambridge University Press, 1979). Also see, J. G. James, "Thomas Paine's Iron Bridge Work, 1785–1803," *Transactions: Newcomen Society* 59 (1987–1988): 189–222; "Iron Arched Bridge Designs in Pre-Revolutionary France," *History of Technology* 4 (1979): 63–69; "Thomas Wilson's Cast-Iron Bridges, 1800–1810," *Transactions: Newcomen Society* 50 (1978–1979): 55–72; "The Eighteenth Dickinson Memorial Lecture: Some Steps in the Evolution of Early Iron Arched Bridge Designs," *Transactions: Newcomen Society* 59 (1987–1988): 153–85.

5. R. J. B. Walker, *Old Westminster Bridge: The Bridge of Fools* (Newton Abbot, England: David and Charles, 1979).
6. *CW*, 2:1052.
7. Ibid., 2:1032.
8. "Diaries of John Hall," no. 13, February 24, 1787, the Library Company of Philadelphia (hereinafter cited as *DJH*). James N. Green, "Report of the Librarian," in *The Annual Report of the Library Company of Philadelphia for the Year 1990* (Philadelphia: The Library Company of Philadelphia, 1991), 8–12.
9. Quoted in Samuel Timmins, "The Industrial History of Birmingham," in *The Resources, Products, and Industrial History of Birmingham and the Midland Hardware District* (Birmingham: The Local Industries Committee of the British Association, 1865), 216. Burke's comment is from a 1777 Parliamentary speech in support of a bill licensing a theater for Birmingham. The world of the toy-maker is explored in Jenny Uglow, *The Lunar Men: Five Friends Whose Curiosity Changed the World* (New York: Farrar, Straus & Giroux, 2002).
10. John Hall to Joseph Capner Lindley, October 24, 1785, in "John Hall Letter Book," p. 31, courtesy of Ms. Polly Munson (transcription by Ms. Munson); Hall to Lindley, December 31, 1785, in ibid., pp. 50 and 53.

Chapter 7: The Magical Iron Arch

1. Peter Kalm, *Travels into North America; Containing Its Natural History and a Circumstantial Account of Its Plantations, and Agriculture in General* (London: T. Lowndes, 1771), 2:177.
2. *DJH*, no. 5, January 14–15, 1786. See also Thomas Paine to Benjamin Franklin, June 6, 1786, in *CW*, 2:1026–28.
3. www.monticello.org/site/research-and-collections/coalbrookdale-iron-bridge; Neil Cossons and Barrie Trinder, *The Iron Bridge: Symbol of the Industrial Revolution*, 2d ed. (Chichester, England: Phillimore, 2002).

4. *DJH*, nos. 8–9, June 16–September 16, 1786.

5. *CW*, 2:1033.

6. *DJH*, no. 11, especially November 27–December 22, 1786.

7. Stevenson Whitcomb Fletcher, *The Philadelphia Society for Promoting Agriculture, 1785–1955*, rev. ed. (Philadelphia: Philadelphia Society for Promoting Agriculture, 1976), 25; *DJH*, no. 10, October 28 and November 8, 1786; Thomas Paine to George Clymer, November 19, 1786, in *CW*, 2:1258.

8. John Hall to Mary Hall Capnerhurst (his sister), January 1, 1787, in "John Hall Letter Book," pp. 157–58, courtesy of Ms. Polly Munson (transcription by Ms. Munson).

9. *DJH*, no. 14, April 20, 1787.

Chapter 8: American Architect

1. Thomas Paine to Benjamin Franklin, June 22, 1787, in *CW*, 2:1262–63; Franklin to the Duc de La Rochefoucauld, April 15, 1787, in *The Papers of Benjamin Franklin*, digital edition by the American Philosophical Society, Yale University and the Packard Humanities Institute, http://www.franklinpapers.org/franklin/framedVolumes.jsp.

2. Marc-Antoine Laugier, *An Essay on Architecture in Which the True Principles Are Explained* (London: T. Obsborne and Shipton, 1755), 238–39.

3. William Chambers, *A Treatise on Civil Architecture* (London: J. Haberkorn, 1759), ii.

4. Quoted in Andrew Saint, *Architect and Engineer: A Study in Sibling Rivalry* (New Haven: Yale University Press, 2007), 297; Antione Picon, *L'Invention de l'ingénieur moderne: l'Ecole des Ponts et Chaussées, 1747–1851* (Paris: Presses de l'École Nationale des Ponts et Chaussées, 1992); and Claude Vacant, *Jean Rodolphe Perronet (1708-1794): "premier ingénieur du roi" et directeur de l'École des Ponts et Chaussées* (Paris: Presses de l'École Nationale des Ponts et Chaussées, 2006).

5. M. Vincent de Montpetit, *Prospectus d'un pont de fer d'une seule arche, proposé, depuis vingt toises jusqu'à cent d'ouverture, pour être jeté sur une grande riviere: présenté au roi le 4 mai 1783* (Paris: Chez l'Auteur, 1783); J. G. James, "Iron Arched Bridge Designs in Pre-Revolutionary France," *History of Technology* 4 (1979): 68; Paine to Franklin, June 22, 1787; Paine to George Clymer, August 15, 1787, in CW, 2:1262–64.

6. Académie des Sciences, Les Procès Verbaux, Folio 333–340, August 29, 1787, J. G. James Papers, Library, Institution of Civil Engineers, London.

7. *Minutes of the First Session of the Thirteenth General Assembly of the Commonwealth of Pennsylvania* (Philadelphia: Hall and Sellers, 1778), 24–29.

8. Paine to Thomas Jefferson, September 9, 1788, in CW, 2:1269; Paine to George Clymer, December 29, 1787, in ibid., 2:1266–67.

9. Ibid., 2:632, 641–42, 650.

10. Ibid., 2:332.

Chapter 9: An Architect and His Patrons

1. On Somerset House, see John Summerson, *Georgian London*, rev. ed. (London: Penguin, 1978). Paine's relationship with Burke is treated in Thomas W. Copeland, *Our Eminent Friend Edmund Burke* (New Haven: Yale University Press, 1949); and Yuval Levin, *The Great Debate: Edmund Burke, Thomas Paine, and the Birth of the Right and Left* (New York: Basic Books, 2014).

2. Edmund Burke to French Laurence, August 18, 1788, in Holden Furber and P. J. Marshall, eds., *The Correspondence of Edmund Burke* (Cambridge: Cambridge University Press, 1965), 5:412; Thomas Paine to Thomas Jefferson, September 9, 1788, in CW, 2:1270.

3. Paine to Burke, August 7, 1788, Library of the American Philosophical Society, Philadelphia.

4. "Specifications of Thomas Paine," *CW*, 2:1031–32; Paine to Thomas Jefferson, September 9, 1788, in ibid., 2:1269.
5. W. H. G. Armytage, "Thomas Paine and the Walkers: An Early Episode of Anglo-American Co-operation," *Pennsylvania History* 18 (January 1951): 16–30; Paine to Jefferson, February 26, 1789, in *CW*, 2:1281.
6. Samuel Walker & Co. Business Journal, (1741–1792), p. 25, Institution of Mechanical Engineers, London; Paine to George Washington, May 31, 1790, in *CW*, 2:1305. On weight, see John Keane, *Tom Paine: A Political Life* (New York: Grove Press, 1995), 281.
7. "Address to Addressers," *CW*, 2:497–98.
8. Paine to Thomas Walker, August, 8, 1790, and Paine to Thomas Walker, September 25, 1790, in Armytage, "Thomas Paine and the Walkers," 24–29.

Chapter 10: The Great Rupture

1. *Public Advertiser* (London), September 15, 1790; Thomas Paine to Thomas Jefferson, September 28, 1790, in *CW*, 2:1315.
2. Edmund Burke, *Reflections on the Revolution in France*, ed. Frank M. Turner (New Haven: Yale University Press, 2003), 66.
3. Quoted in F. B. Lock, *Edmund Burke*, vol. 2, *1784–1797* (Oxford: Oxford University Press, 2006), 339.
4. Burke, *Reflections*, 68, 42.
5. *CW*, 1:279.
6. Ibid., 1:339.
7. Ibid., 1:326–27.
8. Quoted in Mark Philp, "Introduction," in *Thomas Paine: Rights of Man, Common Sense, and Other Political Writings* (Oxford: Oxford University Press, 1995), xxiii.
9. Quoted in E. P. Thompson, *The Making of the English Working Class* (1963; repr., New York: Vintage, 1966), 103.
10. *Public Advertiser*, March 28, 1791; Wyvill quoted in H. T.

Dickinson, "Thomas Paine and His British Critics," *Enlightenment and Dissent* 27 (2011): 27.

Chapter 11: The Specter of Paine

1. *Diary, or Woodfall's Register*, April 11, 1791; *General Evening Post*, April 23–26, 1791. See also *General Evening Post*, April 14–16, 1791.

2. Thomas Paine to Samuel Walker and company, August 30, 1791, Library of the American Philosophical Society, Philadelphia; Paine to John Hall, November 25, 1791, in *CW*, 2:1321.

3. Paine to Messieurs Condorcet, Nicolas de Bonneville, and Lanthenas, June 1791, in *CW*, 2:1315.

4. Ibid., 1:355, 449, 454.

5. Ibid., 1:405, 2:459.

6. Ibid., 2:464.

7. William Theobald Wolfe Tone, ed., *Memoires of Theobald Wolfe Tone* (London, 1827), 2:172–73.

8. Moncure Conway, *The Life of Thomas Paine: With a History of His Literary, Political, and Religious Career in America, France, and England* (New York: G. P. Putnam and Sons, 1908), 1:341.

9. *The European Magazine and London Review* 30 (November 1796): 356; translated from the original Latin.

10. *True Briton* (London), July 29, 1796. On Rowland Burdon, see "The History of Parliament: British Political, Social and Local History" (www.historyofparliamentonline.org/volume/1790-1820/member/burdon-rowland-1757-1838); Maberly Phillips, *A History of Banks, Bankers, and Banking in Northumberland, Durham, and North Yorkshire* (London: E. Wilson, 1894).

11. Thomas Sanderson to Rowland Burdon, January 28, 1793: HO 42/24/119 f. 289–90, National Archives, Kew, England. On the keelmen strikes, see Joseph M. Fewster, *The Keelmen of Tyneside: Labour Organisation and Conflict in the North-*

East Coal Industry, 1600–1830 (Woodbridge, England: Boydell Press, 2011); Edward Raymond Turner, "The Keelmen of Newcastle," *The American Historical Review* 21 (April 1916): 542–45.

12. "Observations and Remarks on the Bridge Proposed to Be Built over the River Wear at or near Sunderland," 79/1/57A, Drawings Archive and Research Library, Sir John Soane Museum, London. For Burdon's friendship with Soane, see Gillian Darley, *John Soane: An Accidental Romantic* (New Haven: Yale University Press, 2000).

13. *A Tour Through the Northern Counties of England, and the Borders of Scotland* (London, 1802), 1:308; Soane Lecture Draft, dated June 1811, MBiii/11/1, Drawings Archive and Research Library, Sir John Soane Museum, London; Edward Cresy, *An Encyclopedia of Civil Engineering, Historical, Theoretical, and Practical* (London: Longman, Brown, Green and Longman, 1847), 1:495.

14. Rev. John Burdon, *Letter to the Wearmouth Bridge Committee* (Sunderland, England: James Williams, 1859), 8, 5; *Sunderland Daily Echo and Shipping Gazette*, September 27, 1879, p. 3. In an indication that the controversy continued, a rebuttal to Picton's claim was published in the *Sunderland Daily Echo* on September 29, 1879, p. 3.

15. *CW*, 2:1054–55; Paine to Thomas Jefferson, October 1, 1800, in ibid., 2:1411.

Chapter 12: Citizen Paine

1. "Reasons for Preserving the Life of Louis Capet" and, "Shall Louis XVI Be Respited," *CW*, 2:554; 556–58.

2. Edmund Burke to Lord Loughborough, January 27, 1793, in P. J. Marshall and John A. Woods, eds., *The Correspondence of Edmund Burke*, vol. 7, *January 1792–August 1794* (Cambridge: Cambridge University Press, 1968), 344.

3. Thomas Paine to Georges-Jacques Danton, May 6, 1793, in *CW*, 2:1335.

4. Anne Cary Morris, ed., *The Diary and Letters of Gouverneur Morris: Minister of the United States to France; Member of the Constitutional Convention, etc.* (New York: Charles Scribner's Sons, 1888), 1:359–60; Paine to Samuel Adams, March 6, 1795, in *CW*, 2:1376.

5. James Monroe to Paine, September 18, 1794, in Thomas Paine, *A Letter to George Washington: on the Subject of the Late Treaty Concluded Between Great Britain and the United States* (London: T. Williams, 1797), 11.

6. R. Barry O'Brien, ed. *The Autobiography of Theobald Wolfe Tone* (London: T. Fisher Unwin, 1893), 2:189.

7. *Morning Post and Gazetteer* (London), October 15, 1798; *The Times* (London), October 16, 1798; *Mirror of the Times* (London), October 13, 1798; *Evening Mail* (London), October 15, 1798; Henry Redhead Yorke, *Letters from France, in 1802* (London: H. D. Symonds, 1804), 2:365–66.

8. *CW*, 2:715.

9. "Of Paine's Letter to Washington," *Massachusetts Mercury* 13 (January 13, 1797), quoted in Simon P. Newman, "Paine, Jefferson, and Revolutionary Radicalism in Early National America," in Simon P. Newman and Peter S. Onuf, eds., *Paine and Jefferson in the Age of Revolutions* (Charlottesville: University of Virginia Press, 2013), 81.

10. *CW*, 1:474, 480.

11. Telford quoted in Samuel Smiles, *Lives of the Engineers* (London: John Murray, 1861), 2:327; Joseph Walker to Thomas Telford, April 14, 1801, and Telford to the Walker brothers, April 21, 1801, both in Thomas Telford Papers, T/LO.23 and T/LO.29, Archives, Institution of Civil Engineers, London.

12. *St. James Chronicle or the British Evening Post* (London), October 23, 1800; Suzanne Francis Brown and Peter Francis, *The Old Iron Bridge: Spanish Town, Jamaica* (Kingston: Fac-

ulty of the Built Environment, Caribbean School of Architecture and Technology, 2005).

Chapter 13: No Nation of Iron Bridges

1. *CW*, 2:1056.

2. *The Republican or, Anti-Democrat* (Baltimore), July 7, 1802, and November 4, 1802; *Virginia Gazette*, November 17, 1802.

3. *New York Evening Post*, November 4, 1802; *Providence Gazette*, November 13, 1802; *New York Gazette*, November 17, 1802.

4. Thomas Paine to Thomas Jefferson, January 12, 1803, in *CW*, 2:1439.

5 Benjamin Rush quoted in Moncure Conway, *The Life of Thomas Paine: With a History of His Literary, Political, and Religious Career in America, France, and England* (New York: G. P. Putnam and Sons, 1908), 2:318; Paine to John Inskeep, February 1806, in *CW*, 2:1480.

6. Charles Willson Peale, *An Essay on Building Wooden Bridges* (Philadelphia: Francis Bailey, 1797), iii. See also Carl W. Condit, *American Building: Materials and Techniques from the Beginning of the Colonial Settlements to the Present*, 2d ed. (Chicago: University of Chicago Press, 1982); Lola Bennett, "From Craft to Science: American Timber Bridges, 1790–1840," *APT Bulletin* 35 (2004): 13–19; George Danko, "The Evolution of the Simple Truss Bridge, 1790–1850: From Empiricism to Scientific Construction" (PhD diss., University of Pennsylvania, 1979); Robert Fletcher and J. P. Snow, "A History of the Development of Wooden Bridges," *Transactions of the American Society of Civil Engineers* 99 (1934): 314–408.

7. Thomas Pope, *A Treatise on Bridge Architecture: In Which the Superior Advantages of the Flying Pendant Lever Bridge Are Fully Proved* (New York: Alexander Niven, 1811), 274–75.

8. Anonymous, "On the Architecture of America" (1790), excerpted in Steven Conn and Max Page, eds., *Building the Nation: Americans Write About Their Architecture, Their Cities, and Their Landscape* (Philadelphia: University of Pennsylvania Press, 2003), 10; Alexis de Tocqueville, *Democracy in America*, ed. J. P. Mayer (Garden City, N.Y.: Anchor, 1969), 468.

9. Palmer quoted in George B. Pease, "Timothy Palmer, Bridge-Builder of the Eighteenth Century," *Essex Institute Historical Collections* 83, no. 2 (April 1947): 104; Owen Biddle, *The Young Carpenter's Assistant. Or, A System of Architecture, Adapted to the Style of Building in the United States* (Philadelphia: Benjamin Johnson, 1805), 52. Angelo Maggi and Nicola Navone, *John Soane and the Wooden Bridges of Switzerland: Architecture and the Culture of Technology, from Palladio to the Grubenmanns* (Mendrisio, Switzerland: Academia di Architettura, 2003).

10. Michel Chevalier, *Society, Manners, and Politics in the United States* (Boston: Weeks, Jordan, and Company, 1839), 271; J. S. Buckingham, *America, Historical, Statistic, and Descriptive* (London: Fisher, Son, & Co., 1841), 2:46; James Fenimore Cooper, *Notions of the Americans: Picked Up by a Travelling Bachelor* (Philadelphia: Lea and Blanchard, 1843), 1:304–5; "Travels of a Tin Pedlar. No. VIII," *The New England Galaxy and United States Literary Advertiser*, January 11, 1828, p. 2; Charles Dickens, *American Notes and Pictures from Italy* (New York: Charles Scribner's Sons, 1898), 168.

11. Frances Anne Butler [Kemble], *Journal of a Residence in America* (Paris: Galignani and Co., 1835), 187; *Public Ledger* (Philadelphia), September 3, 1838; Lee H. Nelson, *The Colossus of 1812: An American Engineering Superlative* (New York: ASCE, 1990).

12. CW, 1:610. Also see Gertrude Himmelfarb, *The Idea of Poverty: England in the Early Industrial Age* (New York: Alfred A.

Knopf, 1983); and Gareth Stedman Jones, *An End to Poverty: A Historical Debate* (New York: Columbia University Press, 2004).

13. *CW*, 1:617.

14. Conway, *The Life of Thomas Paine*, 2:455.

Epilogue

1. *CW*, 1:408–9.

2. Morton J. Horowitz, *The Transformation of American Law, 1780–1860* (Cambridge, Mass.: Harvard University Press, 1977), 112; Andrew M. Schocket, *Founding Corporate Power in Early National Philadelphia* (Dekalb: Northern Illinois University Press, 2007), 71.

3. *Minutes of the Grand Committee of the Whole Convention of the Commonwealth of Pennsylvania* (Philadelphia: Zachariah Poulson, 1790), 89–90; *The Proceedings Relative to Calling the Conventions of 1776 and 1790* . . . (Harrisburg, Pa.: John S. Wiestling, 1825), 138.

Acknowledgments

The staff of the Research Library of Sir John Soane Museum made several summer days spent in the library's collections enormously helpful. I am grateful to the librarian, Dr. Stephanie Coane; Stephen Astley, drawings curator; the head of library services, Susan Palmer; and Bellina Adjei, who expertly handled permissions. At the Pennsylvania State Archives, I benefitted greatly from the expertise of Associate Archivist Michael D. Sherbon, who helped me locate images of the Camelback and Dunlap's Creek bridges.

I was able to review the papers of the late J. G. James, the world's foremost student of early iron bridges, at the Institution of Civil Engineers in London. I am grateful to the institute's librarian, Debbie Francis, for providing me access to these and other invaluable materials.

Ms. Polly Munson of Peyton, Colorado, contacted me some years ago about a brief piece I wrote on Paine for the online journal *Common-Place*. She is a descendant of John Hall and came into possession of his letter books, which she has carefully transcribed and which, to my great benefit, she very generously shared with me.

James N. Green, librarian of the Library Company of Philadelphia, very kindly granted me access to the John Hall Diaries and fielded numerous obscure queries with his customary good humor. Rob Karnowski, a graduate student Florida State University, did yeoman's work, transcribing sections of the Hall Diaries. David Brackett, another descendant of John Hall, shared with me his encyclopedic knowledge of his ancestor's biography. Andrew McCarthy, a brilliant undergraduate at Florida State University, spent a summer expertly checking citations and sharing with me his extensive knowledge of eighteenth-century architecture.

My agent, Lisa Adams of the Garamond Agency, has been an utterly indispensable source of encouragement and wise counsel. At W. W. Norton, I have had the great privilege of working with Steve Forman, a master of the editor's craft. Steve's editorial assistant, Travis Carr, has helped in innumerable ways. I am also grateful for Trent Duffy's exceptionally careful copyediting.

Two friends—Jane Kamensky and Ray Raphael—read an early draft and helped me find my way to the story I was trying to tell.

For everything else, thanks especially to Stacey Rutledge.

Index